Engineering and Entrepreneurship: A Sustainable Approach

Engineering and Entrepreneurship: A Sustainable Approach

Gustav Solomon

NY RESEARCH PRESS
New York

Published by NY Research Press
118-35 Queens Blvd., Suite 400,
Forest Hills, NY 11375, USA
www.nyresearchpress.com

Engineering and Entrepreneurship: A Sustainable Approach
Gustav Solomon

International Standard Book Number: 978-1-64725-413-1 (Hardback)

Cataloging-in-Publication Data

Engineering and entrepreneurship : a sustainable approach / Gustav Solomon.
p. cm.
Includes bibliographical references and index.
ISBN 978-1-64725-413-1
1. Entrepreneurship. 2. Industrial engineering. 3. Technological innovations.
4. Sustainable development. I. Solomon, Gustav.

HB615 .E54 2023
658.421--dc23

Contents

Preface

The main aim of this book is to educate learners and enhance their research focus by presenting diverse topics covering this vast field. This is an advanced book which compiles significant studies by distinguished experts in the area of analysis. This book addresses successive solutions to the challenges arising in the area of application, along with it; the book provides scope for future developments.

Engineering is a field of study that involves the creative application of science, mathematical methods and empirical evidence for developing products, services or processes that help solve some human problem or fulfill a human need. Entrepreneurship refers to the process of creating, launching and operating a new business in expectation of earning a profit from business activities. It is essential for achieving economic growth at domestic and global levels. It involves harnessing the business opportunities in engineering and transforming them into commercially viable and profitable innovations. A technical idea can be turned into a profitable business venture through the implementation of correct accounting practices, administrative understanding, marketing expertise, and reliable commercial operations provided through entrepreneurship. This book discusses the fundamentals as well as modern approaches of engineering and entrepreneurship. It is appropriate for students seeking detailed information in this area as well as for experts.

It was a great honour to edit this book, though there were challenges, as it involved a lot of communication and networking between me and the editorial team. However, the end result was this all-inclusive book covering diverse themes in the field.

Finally, it is important to acknowledge the efforts of the contributors for their excellent chapters, through which a wide variety of issues have been addressed. I would also like to thank my colleagues for their valuable feedback during the making of this book.

Gustav Solomon

1

Sustainable Manufacturing: An Indo-German Perspective

Benjamin Uhlig, Alexander Leiden, Kuldip Singh Sangwan, and Christoph Herrmann

1.1 The History of the Indo-German Center for Sustainable Manufacturing

The global challenge of sustainable engineering needs a global community. Together, the Birla Institute of Technology and Science (BITS) Pilani, and Technische Universität (TU) Braunschweig enable their students and researchers to advance the necessary methods and tools to achieve sustainability goals. Through joint projects in India and Germany, students and researchers have the opportunity to visit another academic environment and experience working in intercultural teams as well as living in a different culture. For both sides, the exchange enables a change of perspective and allows seeing different challenges in sustainable engineering.

The history of this cooperation goes back to the year 2009. From this year on, the team started to apply for funding and began its intense exchange of scientific staff and students. Figure 1.1 shows the development of the exchange over the past ten years—more than 100 students and academic staff has been involved, and additionally, Figs. 1.2 and 1.3 illustrate the milestones of this cooperation.

B. Uhlig (✉) · A. Leiden · C. Herrmann
Chair of Sustainable Manufacturing and Life Cycle Engineering, Institute of Machine Tools and Production Technology (IWF), Technische Universität Braunschweig, Langer Kamp 19b, 38106 Braunschweig, Germany
e-mail: b.uhlig@tu-braunschweig.de

K. S. Sangwan
Department of Mechanical Engineering, Birla Institute of Technology and Science Pilani, Pilani Campus, Pilani 333031, India

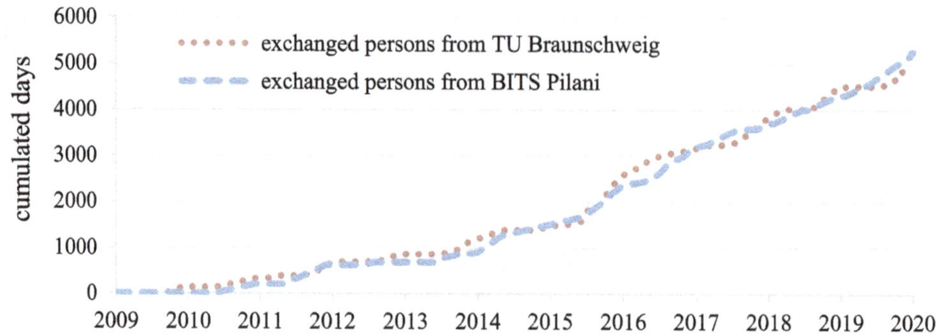

Fig. 1.1 Cumulated days of exchange between TU Braunschweig and BITS Pilani

1.2 Joint Projects and Infrastructure for Research and Education

From the beginning of this cooperation, joint projects are the basis for an exchange, which goes far beyond mutual exchange programs. The projects enable students and researchers to face and tackle future challenges towards a global sustainable development. Besides working on this global challenge, both sides broaden their perspective on different working methods and approaches. In the following, the projects with the greatest effects on society and environment are described in detail.

1.2.1 Lean and Green—Efficiency and Effectiveness in Production

"Lean and Green – Efficiency and Effectiveness in Production" ("Lean & Green") was the first long-term project to be launched within this Indo-German partnership. It started in 2010 and was funded by DAAD's (German Academic Exchange Service) initiative "A New Passage to India" for two years. Within this program, the already existing exchange of students and researchers was intensified.

While the following projects got more specialized, "Lean and Green" provided the basis for further research by creating a common understanding in the topic of sustainable manufacturing. It ought to lead to a better comprehension of the differences and similarities of Indian and German concepts of sustainable manufacturing, of the challenges of adopting and transferring these concepts into new environments, and of the constraints, which occur in this process. Nonetheless, "Lean and Green" was, of course, tracing some major interests. As India's energy requirements increase with rapidly growing Indian economics, a focus of the project was e.g. on how to implement renewable energies, mainly solar energy. This interest rose especially after working with Ghadia Solar in 2009.

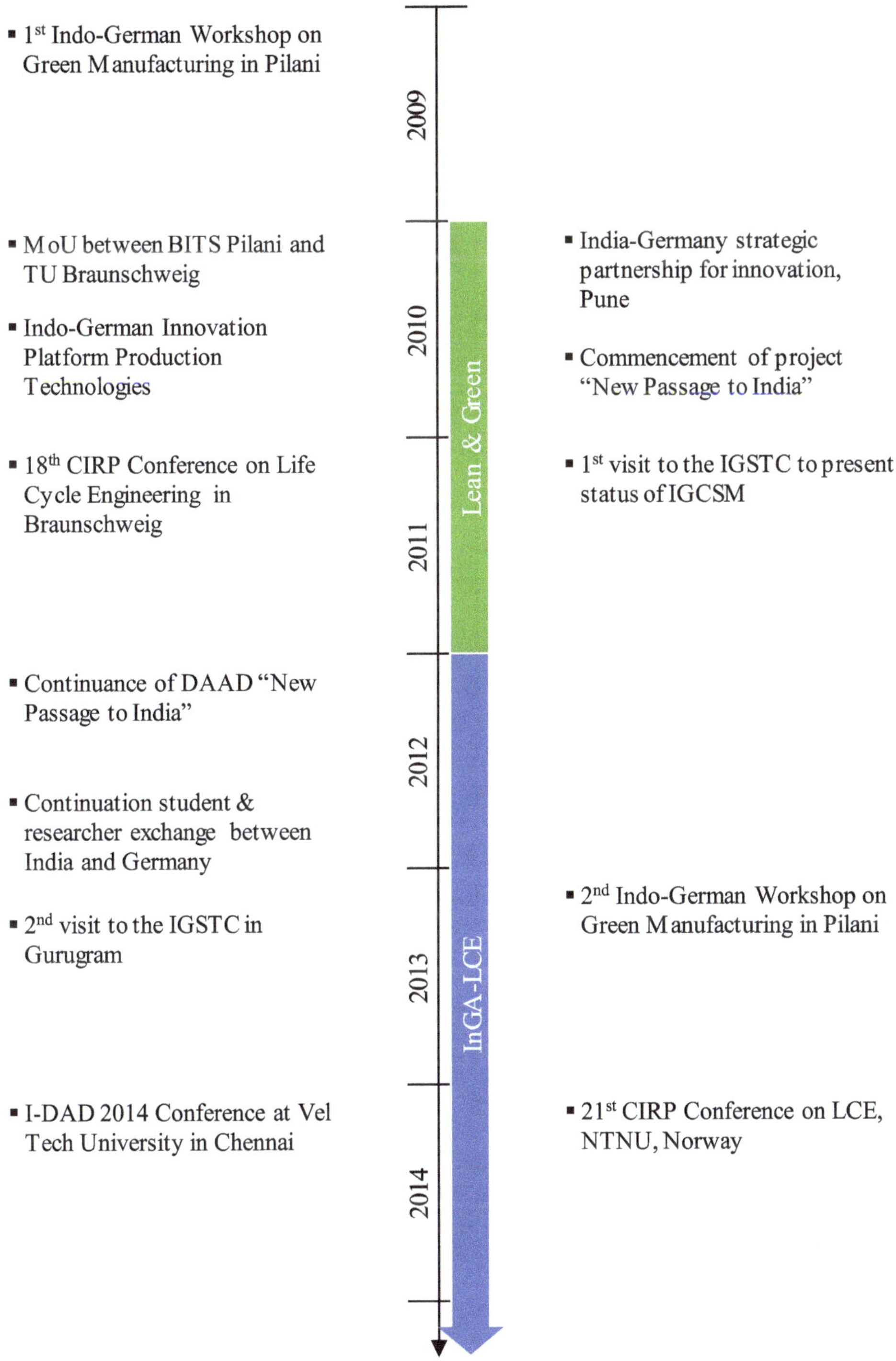

Fig. 1.2 Milestone timeline from 2009 to 2014

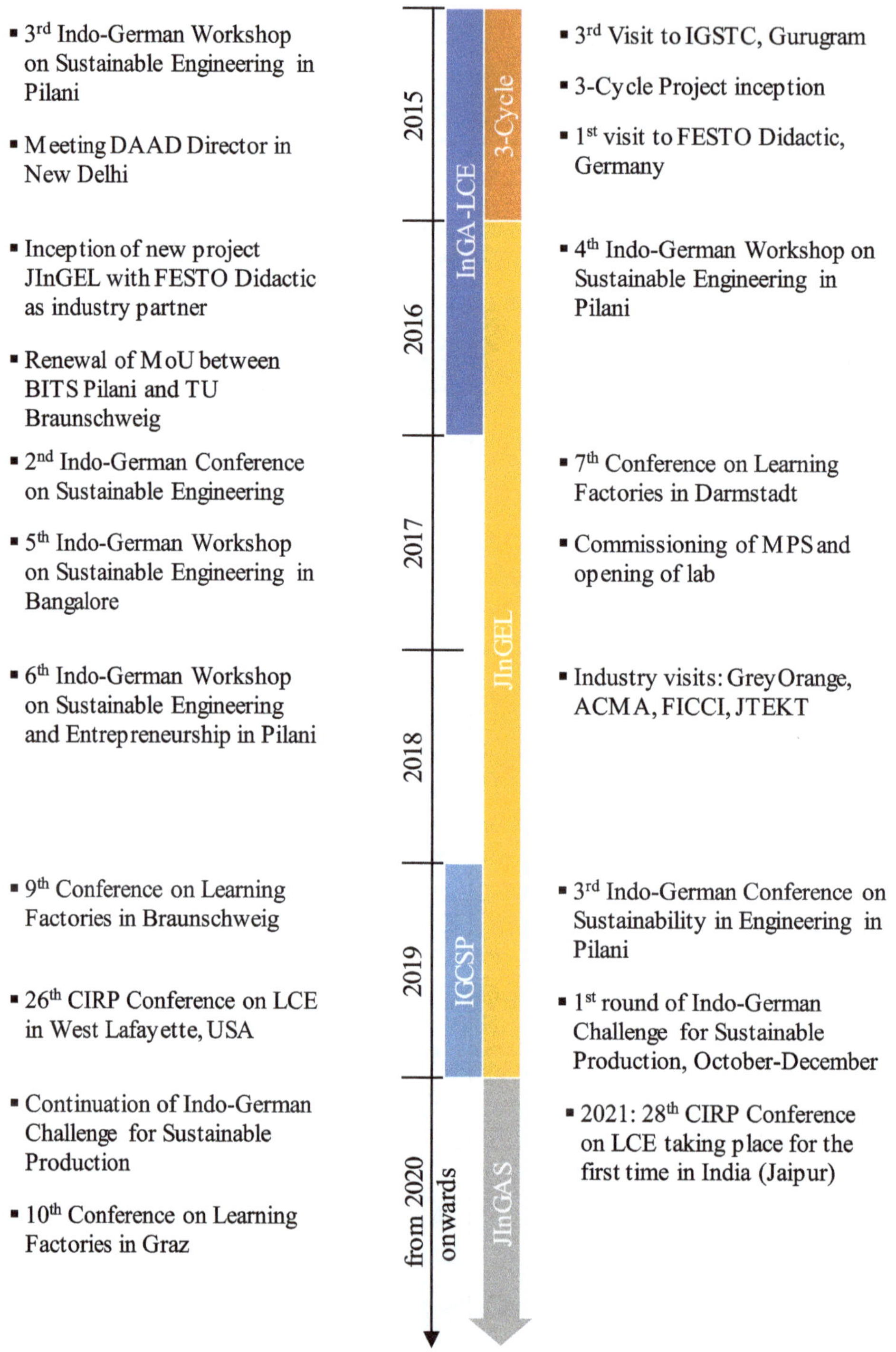

Fig. 1.3 Milestone timeline from 2015 onwards

In general, the project reflected on production systems that were not only economically but also ecologically efficient. Besides an energy-efficient production, the conservation of resources was important. Furthermore, Indian restrictions like availability of qualified employees, capital-intensive equipment, infrastructure, and wage level had to be considered. It was about developing sustainable methods and tools. Technical expertise and principle methods of life cycle engineering had to be adjusted with regard to Indian conditions.

1.2.2 Indo-German Automotive Life Cycle Engineering

Answering to the renewed research call "A New Passage to India" by DAAD, the project "Indo-German Automotive Life Cycle Engineering" ("InGA-LCE") was launched in 2012. With two extensions in 2014 and for a short-term period in 2016, the project lasted in total 4.5 years. It thus has been the longest of the cooperation's projects so far.

It started with the idea to develop a concept of green and electric "Campus Mobility" for BITS Pilani, and aimed, in a broader context, at compiling green transportation systems for India. "Campus Mobility" ought to be an application scenario to think about sustainable, energy-efficient vehicle manufacturing for a country differing from Germany in terms of standards, requirements and infrastructures. India seemed to be the perfect match to the German partners, as it is the country with the world's highest concentration of megacities and quickly increasing mobility needs. The consideration of local particularities of a product's and a vehicle's life cycle, namely different conditions in its development, production, utilization, recycling, was as crucial to the project as the concept of an adaptable vehicle. A comprehensive vehicle concept was to be developed that would work for both of the country-specific mobility scenarios and that could be adapted to the particular characteristics of climate, availability of resources, quality standards, price level, etc. The identification and comparison of two very different mobility concepts constituted the ideal framework for the development of an adaptable vehicle concept, multi-variant, but efficient and effective in production. As the project progressed, the aspect of energy supply became more important. It was crucial to ensure, for instance, that the use of e-mobility would not just shift the problem of CO_2 emissions from traffic to power generation. Starting with a concrete scenario, the project was extended to make the complexity of sustainable mobility concepts visible.

Besides BITS Pilani and IWF (Institute of Machine Tools and Production Technology, TU Braunschweig), the IK (Institute for Engineering Design), and the Automotive Research Association India (ARAI) in Pune were partially involved as they were working on similar topics.

1.2.3 3-Cycle

In recent years, negative environmental impacts due to plastic waste induced by industrial mass production have become visible all over the world and consequently became a global challenge. At the same time, the demand for personalized products, which are manufactured in small batch sizes, has significantly risen and the impact of sustainability on manufacturing companies driven by customer demand and legislation has increased. Subsequently, products made in a more sustainable way can have an economic advantage over their conventional counterparts.

The additive manufacturing technology fused deposition modeling (FDM)—also known as 3D-printing—can be utilized as a production process that can comply with these demands. In FDM 3D-printing a plastic filament is heated and added layer by layer to form a three-dimensional object. With this additive production process, the material is used more efficiently than with conventional subtractive processes, as almost only so much material is added to the part as is actually required.

The production of the basic material for FDM, the filament, has a significant impact on overall costs and environmental impact of the 3D-printed parts. The utilization of filament from plastic scrap material can thus contribute to more sustainable products manufactured with 3D-printing.

In the project "3-Cycle", a small and modular process chain was developed for 3D-printing from extruded waste plastics. Together with researchers and students from BITS Pilani, the process development, application, and evaluation took place both at the Experience Lab at IWF, TU Braunschweig and at BITS in 2015 and 2016. This project was funded by AKB Stiftung.

In the "3-Cycle" project, a great number of students from India and Germany were involved. These students worked individually and in small groups on project tasks mostly during the course of their theses. In total, 16 students and several researchers contributed to the design, construction, testing, and evaluation of the recycling process.

1.2.4 Engineering Education: Joint Indo-German Experience Lab

Engineers do not only participate in the development of products and services, but also in the process of their practical implementation. An education with practical orientation strengthens the competence level of young graduates and provides benefits to industry. Nowadays, the engineering education in Indian universities has a huge potential in combining scientific theoretical education with practical experience. To ease the transition of graduates into the job market, it is advisable to have practical education with industrial context in universities. Especially in reference to management and performing abilities in industrial-like manufacturing environments, shortcomings in current university education have to be compensated. In this context, the idea to adopt the Experience Lab (xLine) of the IWF to the learning environment of BITS Pilani was an ideal extension. The xLine reproduces the most important manufacturing processes in a smaller scale, which makes it possible for students to work in a safe and industrial-like factory environment. The used control systems are the same as in industrial machines and they use the same resources (electricity, compressed air and heat).

Together with the practice partner Festo Didactic SE, which supplied the educational infrastructure, the four-year DAAD project "Joint Indo-German Experience Lab" (JInGEL) was started in 2016. The goal of this project was to transfer the experiences made in the Experience Lab at TU Braunschweig to the Indian university environment. As reference project for further Indian universities and further education institutions, the concept has been introduced at BITS Pilani.

The project offered students and researchers the chance to visit the partner university for a period of three to five months and write their theses within the context of this project. Furthermore, also students and researchers from the department of industrial/organizational and social psychology at TU Braunschweig participated to analyze the effectiveness of the innovative teaching and learning system before and after the integration of the Experience Lab into the BITS curriculum and educational system.

The project aimed at transferring the experiences made in India back to the German academic education. Didactic methods and the curriculum could be enriched by the findings made in Indian teaching and learning culture embedded in their specific economic, ecologic, and social boundary conditions. After a positive evaluation and proof of principle of the Joint Indo-German Experience Lab at the BITS Pilani, the principle will be published as a template to be rolled-out over other BITS campuses and other universities and technical colleges.

1.2.5 Indo-German Challenge for Sustainable Production

Following the call "MINTernational innovativ", a call for supporting projects, which aim at fostering innovative learning and teaching in STEM fields (science, technology, engineering and mathematics), the team established the "Indo-German Challenge for Sustainable Production" (IGCSP). The Club MINTernational has awarded this project as "Best Practice" for initiating new and innovative teaching methods addressing. As part of this project, a joint seminar between BITS Pilani and TU Braunschweig was initiated. The goal of this seminar is to develop industry-relevant solutions in the field of sustainable production. This task had to be fulfilled within a team, which consists of students from TU Braunschweig and BITS Pilani. From October 2019 until December 2019, the first round of IGCSP took place. In this round, the team could also win two apprentices from TU Braunschweig "Gemeinsame Ausbildungswerkstatt" (GAW, joint training workshop). Figure 1.4 shows the overall structure of IGCSP. The major phases can be described as the following.

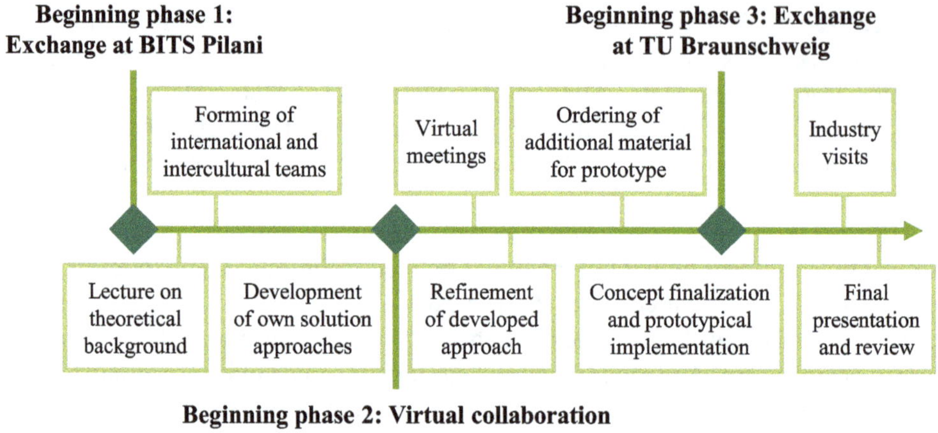

Fig. 1.4 Structure of IGCSP with its three major phases

Phase 1: Exchange at BITS Pilani: Students and teachers from TU Braunschweig and BITS Pilani initially met at BITS Pilani campus in the beginning of October. Here, students gained the required theoretical knowledge during lectures. The lectures have been from the field of Life Cycle Assessment and Sustainable Cyber Physical Production Systems. After that, the students started working in their teams. Each team was working on one specific sub-theme of the seminar. At the end of this phase, a basic procedure of how the problem had to be solved was developed and each of the five groups had created ideas. During this phase, students also had the possibility to gain insights into how industry 4.0 contributes to economic development. Besides this, the participants joint a cross-cultural workshop, in which students, apprentices and teachers jointly developed external image of the other culture and compared these results with the self-image.

Phase 2: Virtual teamwork: After the physical meeting in Pilani, the teams further developed their solution approaches under application of communication software. During their meetings, they discussed how the approach could be implemented in both of the learning factories ("Die Lernfabrik" at TU Braunschweig and "Joint Indo-German Experience Lab" at BITS Pilani). Students also used provided resources such as microcontrollers and sensors. To be able to finalize their prototypes later on in the second phase of exchange, ordering of additional material was possible.

Phase 3: Exchange at TU Braunschweig: In December, the exchange of BITS Pilani students and teachers to TU Braunschweig took place. Here, the teams met for the second time and the students improved and finalized the developed prototype. Then, students presented their approaches to the committee.

1.2.6 Joint Indo-German Academy Towards Sustainability in Engineering, Education and Entrepreneurship

With this project, the teams answers the call "Subject-Related Partnerships with Institutions of Higher Education in Developing Countries" by DAAD. The "Joint Indo-German Academy towards Sustainability in Engineering, Education and Entrepreneurship" (JInGAS) will start in 2020 and end in 2023.

The goal of JInGAS is to successfully continue the 2019 initiated "Indo-German Challenge for Sustainable Production" and thus improving the developed joint seminar in both of the partner universities. It is also intended to support teachers by introducing research-based learning into BITS Pilani curriculum. Besides the field of mechanical engineering, the team will also incorporate other disciplines, namely being education and entrepreneurship. The short-term exchange of the participants of this seminar will be supplemented by long-term exchange of students.

1.3 Joint Workshops and Conferences

Joint research and academic exchange are a major part of this Indo-German cooperation right from the start. To allow the dissemination of research results from the direct partnership's environment to the society, including other universities, industry, NGO and politics, two different formats have been introduced: workshops and conferences.

In September 2009, the **Indo-German Workshop series** started with the 1st Indo-German Workshop in Pilani focusing on measures to increase the energy and resource efficiency in manufacturing companies. Each workshop represented the topics from the current projects and attracted attendees from academia and industry from Germany and India. In 2017, the fifth workshop was jointly organized with Festo India in Bangalore to reach partners from the other side of the subcontinent. One year later, in 2018, the 6th Indo-German Workshop on Sustainable Manufacturing and Entrepreneurship took place in Pilani.

The second format are conferences. As the cooperation heavily grew in recent years, the team started organizing and hosting a **series of academic Indo-German conferences**. Researchers of other institutions from Germany and India joined these series to present their results to foster the topics of the cooperation. The most recent conference was the 3rd Indo-German Conference on Sustainability in Engineering, Education and Entrepreneurship: Enhancing Future Skills and Entrepreneurship. Besides the presentation of the authors' contributions, industry and NGO, partners presented inspiring insights into their work. The contributions follow this chapter of this book.

Right from the beginning, the partners jointly participate in international conferences to present the outcomes of their research. The major two have been the CIRP[1] Conference on Life Cycle Engineering and the conference on Learning Factories (CIRP sponsored). The team actively participates in these conferences and publishes joint research outcomes. On CIRP Conference on Life Cycle Engineering 2011 in Braunschweig, the team had four joint contributions (Herrmann et al. 2011; Jindal and Sangwan 2011; Mittal and Sangwan 2011; Sangwan 2011). On CIRP Conference on Learning Factories 2017 the team published outcomes from engineering education research (Büth et al. 2017; Juraschek et al. 2017). In addition, other conferences in Asia were used to widespread the idea of learning factories (Leiden et al. 2018). One year later 2018, a study on solar energy on campus has been presented (Sangwan et al. 2018). In 2019 CIRP Conference on Life Cycle Engineering, one joint paper on environmental impact of machining (Sihag et al. 2019) has been published. During Conference on Learning Factories, the implementation of cooling processes into learning factories (Vogt et al. 2019) and a study about research-based learning for skill development (Singh et al. 2019) were presented. Further joint publications can be obtained from the references section (Herrmann et al. 2009; Halubek et al. 2010; Mittal et al. 2012; Bhakar et al. 2013; Mittal et al. 2013; Bhakar et al. 2015; Juraschek et al. 2016).

As an outlook, the team is very grateful to announce that the 28th CIRP Conference on Life Cycle Engineering will take place in Jaipur, India (see www.lce2021.in). This will be the first time that such a prestigious CIRP conference is going to be held in India.

1.4 Sponsors and Supporters

From the very beginning of this cooperation, intensive support from different institutions has been realized. Without this support, the cooperation would not be as successful as it is today. Besides the personal efforts of all involved people, the team is very grateful to the sponsors and supporters.

Most importantly, the team wishes to thank DAAD (German Academic Exchange Service) for the long-term support. Without their financial support, it would not have been possible to establish such a successful and intense partnership. DAAD also made it possible to publish this book.

Secondly, the team thanks Festo Didactic SE to support the infrastructure for "JInGEL" at BITS Pilani. The constant positive feedback from students and teachers motivates to foster even more the integration of their modules into BITS Pilani curriculum.

In addition, AKB Stiftung, Stifterverband, the Daimler-Fonds, the Daimler and Benz Foundation have been important sponsors of this cooperation. They made it

[1]CIRP: International Academy for Production Engineering.

possible to realize the innovative projects "3-Cycle" and "Indo-German Challenge for Sustainable Production".

Moreover, Rajni Singh, Luise Mayer, Rishi Kumar, Vikrant Bhakar, Stefan Böhme, Christina Brand, Lennart Büth, Kailash Chaudhary, Max Juraschek, Gerrit Posselt, Linda Sandrock, Narpat Ram Sangwan, Anne-Marie Schlake, Nitesh Sihag, have made important contributions to these proceedings. The team is very thankful to all of them.

Last but not least, the team wishes to express tremendous gratitude to all students, researchers, institutions and partners that were part or supported our cooperation. Their personal involvement and motivation, but also patience and understanding, have been the foundation of our cooperation.

References

Bhakar V, Purwar A, Nayak A, Kumar K, Singer J, Egede P, Herrmann C, Sangwan KS (2013) Life cycle assessment of lathe processes for training application. Presented at the International conference on management and business innovation, Jaipur

Bhakar V, Sihag N, Gieschen R, Andrew S, Herrmann C, Sangwan KS (2015) Environmental impact analysis of a water supply system: study of an Indian University Campus. Procedia CIRP 29:468–473. https://doi.org/10.1016/j.procir.2015.02.061

Büth L, Bhakar V, Sihag N, Posselt G, Böhme S, Sangwan KS, Herrmann C (2017) Bridging the qualification gap between academia and industry in India. Procedia Manufact 9:275–282. https://doi.org/10.1016/j.promfg.2017.04.009

Halubek P, Kassel T, Kücükay F, Herrmann C (2010) Product quality and production costs in automotive industry—a simulation based evaluation of product- and quality strategies. Presented at the 1st International conference on automotive materials and manufacturing, Pune, India

Herrmann C, Thiede S, Kuntzky K, Böhme S, Frauenhofer M, Gadhia D (2009) Indian solar thermal technology: potentials and challenges. In: Shunmugam MS, Ramesh Babu N (eds) Sustainable product development and life cycle engineering: proceedings of the 7th global conference on sustainable manufacturing, 2009, Chennai, India. Indian Institute of Technology, Madras, 2009

Herrmann C, Sangwan KS, Mennenga M, Halubek P, Egede P (2011) Assessment of alternative propulsion systems for vehicles. In: Hesselbach J, Herrmann C (eds) Glocalized solutions for sustainability in manufacturing. Springer, Berlin, Heidelberg, pp 51–56

Jindal A, Sangwan KS (2011) Development of an interpretive structural model of barriers to reverse logistics implementation in Indian industry. In: Hesselbach J, Herrmann C (eds) Glocalized solutions for sustainability in manufacturing. Springer, Berlin, Heidelberg, pp 448–453

Juraschek M, Bhakar V, Posselt G, Kuluza A, Böhme S, Herrmann C, Sangwan KS (2016) Recycling of plastic waste using fused deposition modeling as a production process. Presented at the Indo German conference on sustainability 2016, Chennai

Juraschek M, Cerdas F, Posselt G, Herrmann C (2017) Experiencing closed loop manufacturing in a learning environment. Procedia Manufact 9:57–64. https://doi.org/10.1016/j.promfg.2017.04.046

Leiden A, Posselt G, Bhakar V, Singh R, Sangwan KS, Herrmann C (2018) Transferring experience labs for production engineering students to universities in newly industrialized countries. In: IOP conference series: materials science and engineering, p 012053

Mittal VK, Sangwan KS (2011) Development of an interpretive structural model of obstacles to environmentally conscious technology adoption in Indian industry. In: Hesselbach J, Herrmann C (eds) Glocalized solutions for sustainability in manufacturing. Springer, Berlin, Heidelberg, pp 383–388

Mittal VK, Sangwan KS, Herrmann C, Egede P, Wulbusch C (2012) Drivers and barriers of environmentally conscious manufacturing: a comparative study of Indian and German organizations. In: Dornfeld DA, Linke BS (eds) Leveraging technology for a sustainable world. Springer, Berlin, Heidelberg, pp 97–102

Mittal VK, Egede P, Herrmann C, Sangwan KS (2013) Comparison of drivers and barriers to green manufacturing: a case of India and Germany. In: Re-engineering manufacturing for sustainability, pp 723–728

Sangwan KS (2011) Quantitative and qualitative benefits of green manufacturing: an empirical study of Indian small and medium enterprises. In: Hesselbach J, Herrmann C (eds) Glocalized solutions for sustainability in manufacturing. Springer, Berlin, Heidelberg, pp 371–376

Sangwan KS, Herrmann C, Soni MS, Jakhar S, Posselt G, Sihag N, Bhakar V (2018) Comparative analysis for solar energy based learning factory: case study for TU Braunschweig and BITS Pilani. Procedia CIRP 69:407–411. https://doi.org/10.1016/j.procir.2017.11.018

Sihag N, Leiden A, Bhakar V, Thiede S, Sangwan KS, Herrmann C (2019) The influence of manufacturing plant site selection on environmental impact of machining processes. Procedia CIRP 80:186–191. https://doi.org/10.1016/j.procir.2019.01.023

Singh R, Devika, Herrmann C, Thiede S, Sangwan KS (2019) Research-based learning for skill development of engineering graduates: an empirical study. Procedia Manufact 31:323–329. https://doi.org/10.1016/j.promfg.2019.03.051

Vogt M, Uhlig B, Sangwan KS, Herrmann C, Thiede S (2019) Implementation of a cyber-physical cooling storage station in a learning factory. Procedia Manufact 31:142–147. https://doi.org/10.1016/j.promfg.2019.03.022

2

Sustainable Environment and Oxygen-Enriched Combustion

Ankur Mittal, Ashutosh Saxena, and Bibekananda Mohapatra

Abstract There is an absolute need to adopt innovative technologies to improve energy efficiency with minimum possible environment emission and conservation of natural resources. Oxygen-enriched combustion is one of the latest technologies that may improve combustion efficiency depending on the exhaust gas temperature and percentage of oxygen in the combustion air. Cement industry is responsible for approximately 8% of the global anthropogenic CO_2 emissions (IPCC, 2006) and the cement market is expected to grow with increased industrialization and urbanization. In typical cement manufacturing process, 60% of CO_2 emissions are due to the transformation of limestone to lime (the calcination process) and rest 40% is due to fuel combustion in pyro processing. The air is used as an oxidizing agent content in industrial combustion processes that has maximum nitrogen component (78–79%) by volume. The chemically inert nitrogen dilutes the reactive oxygen and carries away some of the energy in the hot combustion exhaust gas during the air-fuel combustion process. An increase in oxygen in the combustion air can reduce the energy loss in the exhaust gases and increase the fuel combustion efficiency. Oxygen enrichment is helpful in curbing gaseous emission. By increasing oxygen content in air, N_2 content is limited that leads to less NOx in exhaust gases. In this condition exhaust gases are more CO_2 rich that are partially recirculate along with combustion air. In CO_2 rich exhaust gases, water vapour is removed though condensation process and remaining CO_2 is captured through CCS technology.

Keywords Oxygen enrichment · CO_2 emission · Combustion

A. Mittal (✉) · A. Saxena · B. Mohapatra
Centre for Mining, Environment, Plant Engineering, & Operation, National Council for Cement and Building Materials, Ballabgarh 121004, India
e-mail: ankurmittal82@gmail.com

2.1 Use of Oxygen Enriched Air to Improve Combustion Processes

Combustion is a chemical process in which a substance reacts rapidly with oxygen and releases heat as product. The original substance is called the fuel and the source of oxygen is called the oxidizer (Mathieu 2006).

Oxygen is required for any combustion process and ambient air is the most common source of oxygen that contains about 79% Nitrogen by volume. N_2 is inert gas and does not contribute in heat released through combustion reaction (Oates 1998). The nitrogen contained in air actually inhibits fuel from reacting with oxygen. This results in a flame temperature below that attainable with pure oxygen (Schorcht et al. 2013). Oxygen enrichment which is known as increased O_2% in combustion air. This improves the overall combustion process and the resulting heat transfer increases flame temperature and the amount of available heat (Eriksson 2015). Oxygen enrichment process enhances burning zone control and improves kiln stability in the industrial furnace or rotary kiln (Eriksson et al. 2014). We get more consistent kiln operation, better clinker quality, and increased production or alternative fuel substitution rate among all possible positive results (Liu et al. 2015). Oxygen is added to combustion air to increase specific fuel rates in kg fuel/kg air (supplemental enrichment) or reduce overall air volume (equivalent enrichment) (Sharma et al. 2017). Overall gas flow rates are reduced and thermal efficiency increases by substituting pure oxygen either for a portion or total combustion air (Gao et al. 2017). For an example 21 m^3/h of pure oxygen can replace 100 m^3/h of air, thereby reducing the total flue gas volume by 79 m^3/h. The benefits of oxygen enrichment can be achieved even at very low levels of enrichment (Hokfors 2014).

The volumetric reduction in exhaust gases is easily illustrated by comparing the combustion reactions of air/methane (2.1) and oxygen/methane (2.2). Similar reductions in combustion products occur for all fuels due to the elimination or reduction of nitrogen contained in air.

$$CH_4 + 2\,O_2 + 7.5\,N_2 = 2\,H_2O + CO_2 + 7.5\,N_2 \tag{2.1}$$

$$CH_4 + 2\,O_2 = 2\,H_2O + CO_2 \tag{2.2}$$

On the above, it can be easily understood that for the air/methane reaction, there are 10.5 volumes of combustion products, compared to only three volumes of combustion products for the oxygen/methane flame. The adiabatic flame temperature of the oxygen/methane flame is roughly 800 °C higher than the air/methane flame due to the elimination of nitrogen.

2.2 Calcination Process in Rotary Cement Kiln

Lime (calcium oxide, CaO) is produced by calcination of limestone, containing a high concentration of calcium carbonate ($CaCO_3$) better known as Limestone. Limestone is an abundant natural raw material where lime is used for environmental purposes like waste neutralization or flue gas desulphurization. Limestone is also widely used in many industrial processes like in formation of metallurgical Slags or for production of paper pigments. The method used is based on multi-component chemical equilibrium calculations to predict process conditions.

Lime is produced by calcination of calcium carbonates in industrial kilns. The mineral calcite containing the calcium carbonates is the main component in naturally abundant limestone. The limestone is quarried or mined, mechanically pre-treated and delivered to the lime plant. One of the most common kiln types is the rotary kiln. Calcination is an endothermic reaction requiring heat to evolve gaseous carbon dioxide from the calcite to form lime (Fig. 2.1).

$$\underset{(1.00\,g)}{CaCO_3} \xrightarrow[1{:}783\,kJ]{\Delta} \underset{(0.56\,g)}{CaO(s)} + \underset{(0.44\,g)}{CO_2(g)} \tag{3.3}$$

The calcination starts between 800 and 900 °C and the operational solid temperature usually reaches 1000–1200 °C. The calcination temperature is dependent on the partial pressure of carbon dioxide in the kiln.

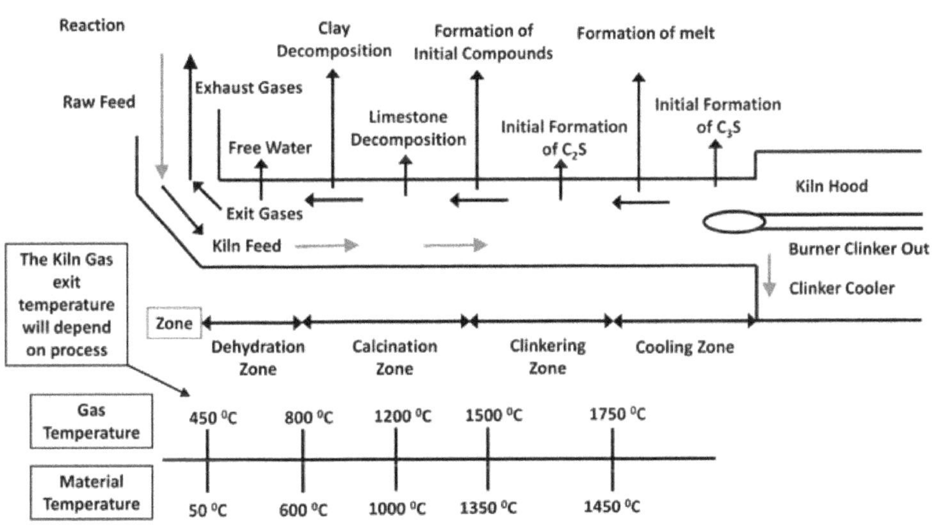

Fig. 2.1 Illustration of calcination/combustion process in kiln

2.3 Oxygen Injection Methods in Rotary Cement Kiln

The two main oxygen enrichment methods for the kiln burner are either by general enrichment or focused enrichment. General enrichment is a method of adding the oxygen to combustion air piping (typically primary air) to increase the percentage of oxygen above 21%. This method is simple to retrofit and is an inexpensive way to obtain some of the benefits of oxygen enrichment.

Second method is focused enrichment of the kiln burner. This is accomplished by adding lance(s), rather within the burner or adjacent to the burner for injecting oxygen into the flame. This method of enrichment provides the most effective use of oxygen for increasing production or alternative fuel utilization. This is related with modifying the flame heat release profile. Special attention must be given to burner flame shape to maximize performance and to avoid degradation of the protective coating on the kiln refractory. Often various techniques can be used to allow the producer to adjust the flame length and heat release pattern to optimize the overall performance, economics and emissions. Proper lance design along with the evaluation of burner primary airflow is essential to ensure successful implementation.

CFD has proven to be a valuable tool when evaluating the proper method of oxygen enrichment for predicting situations that may occur when using oxyfuel combustion technology in a rotary cement kiln. Computational Fluid Dynamics (CFD) methodology can be used to investigate the temperature distribution, burn out rate of coal, flame shape, release of NOx etc. The result indicates significant temperature increase in the rotary kiln. Coal particles combust more rapidly near the burner and final degree of burnout also increases (Fig. 2.2).

It may be clearly noticed from figures that the highest temperature zone around the core of the flame has been increased while the temperature at the walls of the kiln has remained similar to the conventional air combustion flame case. This can translate into increased production, increased alternative fuel usage and reduced emissions.

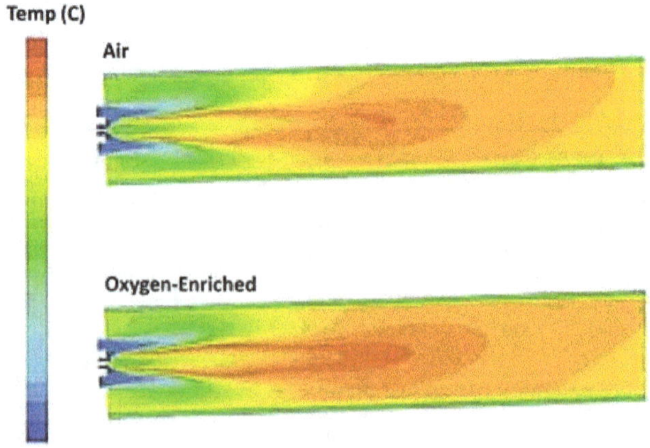

Fig. 2.2 Flame profile with and without oxygen enrichment

2.4 Result Analysis and Research Gap

Most cement plants adopt ambient air for combustion purpose that is catered through induced draft fans. Nowadays plants are pondering over exploring oxygen enrichment at their facility for sake of increasing productivity as well as improving fuel combustion efficiency. The approach for predicting the required oxygen usage to affect desired changes in kiln production and firing rate relies on the concept of heat availability. For a given amount of fuel and a constant excess oxygen level, addition of oxygen will increase the percentage of high grade heat. This is useful because clinker production is limited by the high-grade heat while excess low grade heat simply goes up the stack. Some results in rotary kiln with varying oxygen % are presented below (Tables 2.1 and 2.2).

Despite of encouraging results as above, some concerns are there for providing a early and significant thought like:

1. Cost of air separation technology must be economically feasible with a lucrative low pay back period.
2. The lack of extensive experiments and for extended periods of time is still a concern for easy adoption of this technology.
3. Higher flame temperatures with oxygen enrichment can cause overheating in the burning zone leading to a depletion of kiln coating and excess refractory wear.

2.5 Summary

Oxygen-enriched combustion technology can improve the fuel combustion conditions of cement production, increase flame temperature, shorten the time needed for combustion and achieve complete combustion enabling cement plants to increase the flame radiation heat ability of a material and improving the whole system's thermal

Table 2.1 Influence on flame temperature inside the kiln

Oxygen concentration in primary air (%)	Flame temperature (°C)	Clinker temperature (°C)
23.0	1381	1202
24.0	1322	1312
25.4	1431	1378
25.8	1435	1335

Table 2.2 Influence on energy consumption of clinker sintering

Oxygen concentration in primary air (%)	Total coal consumption (t/h)	Clinker sintering energy consumption (kcal/kg clinker)
23.0–25.0	31.2	663.1
25.1–27.0	29.8	660.8

efficiency. It can also reduce waste gas and dust and harmful gas emissions, which is beneficial to energy conservation and emissions reduction. It can improve the production efficiency and quality of cement. Therefore, applying oxygen-enriched combustion technology in cement production can yield economic, social and environmental benefits.

- Oxygen-enriched combustion can improve furnaces flame temperatures by 100–300 °C without any increase in fuel. This step is resulted into remarkable energy-saving effect.
- Flame temperature can rise by 200–300 °C when oxygen content increases by 4–5%.
- Oxygen enriched combustion increases in the whole furnace temperature and the furnace-heated material is easier to heat that increases thermal efficiency.
- Oxygen enrichment improves the combustion of pulverized coal, so that the kiln burns more fully. The kiln temperature improves with improvement in coal combustion rate.
- Free lime (F CaO) content in cement clinker decreases, stability and strength increase, better quality clinker results and the cement quality improves with improved combustion in the rotary kiln.
- Combustion is completed with less amount of air due to excess oxygen content in combustion air.
- There is less smoke production with reduction in dust pollution and environmental pollution. The successful application of oxygen-enriched combustion technology in the rotary kiln cement production industry will bring huge economic and social benefit aspects.

The oxy-fuel combustion process involves burning of pulverized coal in an oxygen enriched atmosphere that consists of pure O_2 that is mixed with recycled flue gas. This process differs from the conventional fuel combustion process where ambient air serves as the only oxidant. This entails specific conditions regarding thermo-physical properties, which affect both combustion characteristics and heat transfer.

Apart from efficient fuel combustion, oxyfuel combustion technology equally provides an opportunity to simplify carbon dioxide (CO_2) capture in coal fired cement plants. The capital cost, energy consumption, and operational challenges of oxygen separation are a primary challenge of cost-competitive oxy-combustion systems. Oxy-combustion system performance can be improved by two means:

- By lowering the cost of oxygen supplied to the system and
- By increasing the overall system efficiency.

The R&D within the advanced combustion systems program is aimed at strategies to improve oxy-combustion system efficiency and reduce capital cost, offsetting the challenges of oxygen production.

References

Eriksson M (2015) Sustainability measures in quicklime and cement clinker production. Doctoral dissertation, Umeå universitet

Eriksson M, Hökfors B, Backman R (2014) Oxyfuel combustion in rotary kiln lime production. Energy Sci Eng 2(4):204–215

Gao T, Shen L, Shen M, Liu L, Chen F, Gao L (2017) Evolution and projection of CO_2 emissions for China's cement industry from 1980 to 2020. Renew Sustain Energy Rev 74:522–537

Hokfors B (2014) Phase chemistry in process models for cement clinker and lime production. Doctoral dissertation, Umeå universitet

Liu YQ, Zhang AM, Qing S, Li FS, Yang SP, Yang ZF (2015) Experimental study on improving cement quality with oxygen-enriched combustion technology. IOP Conf Ser Mater Sci Eng 103(1)

Mathieu P (2006) The IPCC special report on carbon dioxide capture and storage

Oates JAH (1998) Lime and limestones: chemistry and technology, production and uses. Wiley-VCH, Weinheim

Schorcht F, Kourti I, Scalet BM, Roudier S, Sancho LD (2013) Best Available Techniques (BAT) Reference document for the production of cement, lime and magnesium oxide. European commission joint research centre institute for prospective technological studies (Report EUR 26129 EN). Publications Office of the European Union, Luxembourg

Sharma et al (2017) Industrial process heating—technology assessment. J Clean Prod 8(1):119–130

3

Optimizing a Semi-Process Industry: Implementation of Lean Manufacturing

Jaiprakash Bhamu, Jagdish Bhadu, and Kuldip Singh Sangwan

Abstract This paper aims at proving that lean manufacturing can be successfully implemented in a semi-process industry, to reduce defects, waste, waiting time, processing time, inventory, and space. A case study was performed on a medium sized ceramic industry which resulted in waiting time reduction by 32%, processing time reduction by 1.02%, reduction in inventory by 36%, defects reduction (average) by 40.51%, reduction in manpower by 6.85% and cost saving of US $ 0.61 million in one financial year. It might represent new opportunities for lean practitioners/researchers who are interested in substantiating their lean endeavors in semi-process industries.

Keywords Lean manufacturing · Waste reductions · Value stream mapping

3.1 Introduction

The goal of lean manufacturing (LM) is to become highly responsive to customer demand by reducing the waste in human efforts, inventory, time to market and manufacturing space while producing quality products effectively and efficiently (Bhamu et al. 2012). The LM implementation was started with automobile industry and soon was adopted by other industries including textile, construction, medical, food, electrical & electronics, furniture, and services (Bhamu and Sangwan 2014). The combined application of Value Stream Mapping tools (VSM and Kanban) has shown the waste reduction in textile enterprise and increase in inventory turnover that affects

J. Bhamu (✉) · J. Bhadu
Department of Mechanical Engineering, Engineering College Bikaner, Bikaner 334004, RJ, India
e-mail: bhamujp@gmail.com

J. Bhadu
e-mail: erj.bhadu@gmail.com

K. S. Sangwan
Department of Mechanical Engineering, Birla Institute of Technology and Science, Pilani-Campus, Pilani 333031, RJ, India

the operational profitability. Low inventory in terms of free space provides opportunities to expand production by adding more machinery (Carvalho et al. 2018). Though VSM has many benefits, but when applied mistakenly, it may generate poor results that lead to bad decisions, both technically and financially. The future opportunities for VSM implementation, especially with regard to increased productivity and reliability of this lean tool were also discussed (Yuvamitra et al. 2017). VSM has the capability to diagnose a system with a lean perspective and to identify the opportunities for various lean projects/tools. The author reported the improvement plan, highlighting key lean projects/tools to perform for achieving their vision (Sangwa and Sangwan 2018). Ongoing reforms in the power sector and expansion of distribution infrastructure has resulted in demand for insulators, especially high-tension insulators (33 kV and above). The Indian share of the USD three billion in global electric insulator market is expected to reach 25-30% by the end of 2019 at the compounded annual growth rate of 6.4% (Rother and Shook 1999).

The LM has been adopted by all types of manufacturing systems; however, the implementation of lean manufacturing in the continuous process industry has been less partly because of certain difficulties in the implementation in these type of industries (Jimenez et al. 2011). The application of lean manufacturing in ceramic industries (a semi-process industry) is challenging, as first half of the process is continuous type and the second half is discrete part manufacturing. The defective ceramic products cannot be reworked and recycled, and hence pose great threat to the environment. In addition, it leads to the loss of resources and missed opportunity cost. Therefore, the purpose of present work is to show the applicability of lean manufacturing in the ceramic sector through reduction of defects, waste, waiting time, processing time, inventory, rejections, and area required.

3.2 Selection of the Case Company

The case study was carried out at the company called XXX (a semi-process industry), is manufacturing high voltage and extra high voltage alumina porcelain insulators. Though it is challenging to implement LM in the industry, but it was selected as the top management was willing to implement lean manufacturing to improve productivity, quality and flexibility.

The production process was understood and a process sequence is created as shown in Fig. 3.1. The sequence of operation includes pre- kiln, kiln and post-kiln processes (Narapinji 2016). Tests are carried out before final inspection, which includes parallelism, eccentricity, visual defects, full dimension, and production specific requirement tests. The quality control passed insulators are packed in break proof wooden packs. The last process is loading and dispatching by the marketing department.

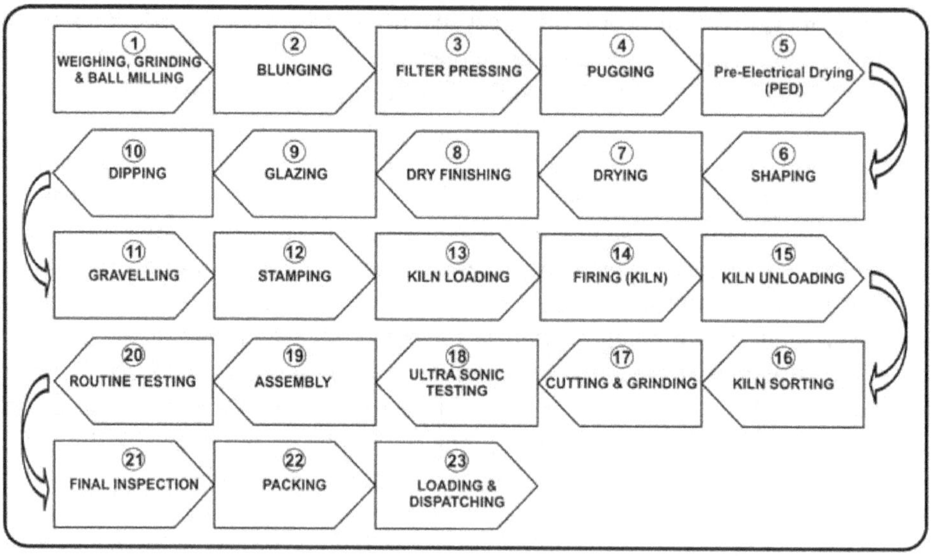

Fig. 3.1 Production process for the solid electric insulators

3.3 Preparation and Analysis of Current Value Stream Map

Current Value Stream Mapping (CVSM) serves as a starting point to help every person in the manufacturing system to identify waste. Value stream perspective means working on a bigger picture and trying to improve the whole process, not just individual processes (Fig. 3.2). There are two inventories mentioned below the inventory triangle in the current state map in which top data represent the inventory of solid core and below one indicates the inventories of all types of insulators. The following observations and analyses can be made from the current state of the organization:

- The processing time is approximately 13 days (18702.07 min), while the waiting time is about 24.84 days. Apart from this, a large amount of inventory is piled up before drying, kiln sorting, cutting & grinding, assembly, and final inspection processes/workstations.
- The inventory of 90 days locked at incoming material stores for the fear of unavailability of raw material.
- There is a large work-in-process (WIP) before major processes/workstations due to fear of unavailability of right quantity of finished goods. This also causes excess waiting time (24.84 days) and results in higher lead time.

Waste analysis is carried out for each process to find out the major waste, their causes and solutions. All the improvement initiatives/kaizens implemented to improve the processes have been discussed in the Sect. 5 of actions, results and discussion.

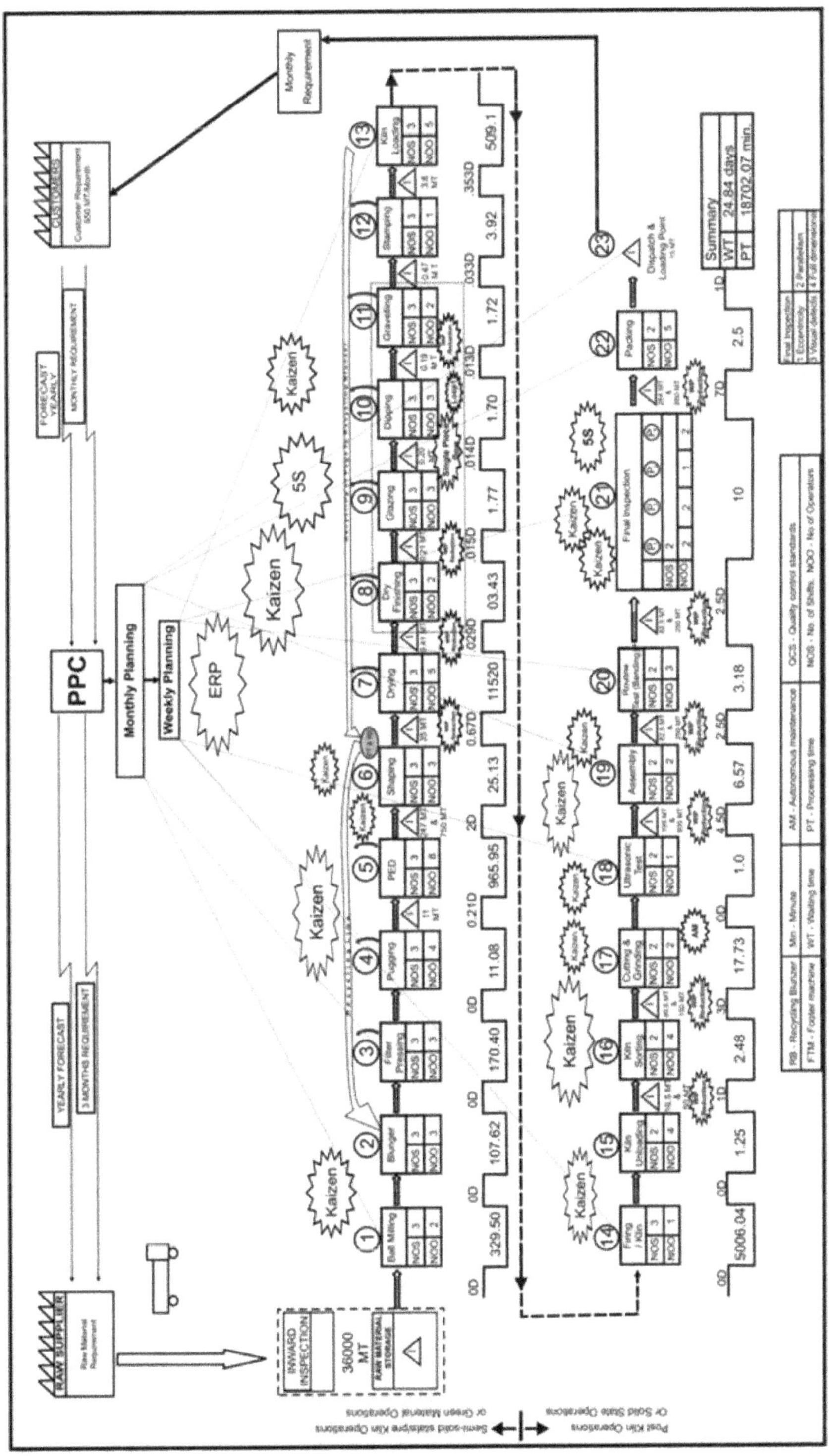

Fig. 3.2 Current value stream map

3.4 Actions, Results and Discussion

Based on the root causes identified in the previous section, specific kaizens were implemented as shown in Fig. 3.1 to reduce various types of wastes. Following general actions were also taken to improve productivity and reduce waste:

- The machine operators were trained to be multi-skilled with objective to start autonomous maintenance by the operators rather than the maintenance staff. It was started in three sections, cutting and grinding, shaping, and testing section.
- The importance of human resources in the lean implementation is very crucial. Shop floor supervisors were extensively trained to identify bottlenecks and take corrective actions to reduce the scrap and rework.
- Regular seminars and workshops were organized by external experts to improve communication among employees.

Some of the performance parameter improvements achieved are as follows:

3.4.1 Processing Time

The implementation of the kaizens and reduction in wastes has resulted in decreasing the processing time from 18,702.07 to 18,510.51 min, i.e. a reduction of 191.56 min. Some of the implemented kaizens are: a rack with hook, knife, spanner, hopper and drainage pipe was placed between two ball mills (Process time (PT)) saved by 0.5 min); the charging process has been improved by charging the ball mill simultaneously with raw materials and water (activity combination) (PT saved by 18 min, energy saved); a digital clock and a siren were installed on each filter press to ensure it does not over-run (average PT saved by 20 min, defect reduction); installation of electric powered bolt tighteners on every bending machine led to PT time saving (PT saved by 0.81 min); the location of oil tanks were changed to bring them at their point of use (POU) for oiling of pug ends (PT saved by 10 min); permanent attachment were added to ball mills for unloading of slurry (PT saved by 13 min); application of wire mesh and covering both the sides of the insulator with polythene started before the pugs were loaded in the PED (PT saved by 60 min); improved insulator cleaning method (PT saved by 30 min).

3.4.2 Waiting Time and Work-in-Process Reduction

Reduction in defects resulted in reduction of raw material inventory and WIP inventory at many workstations. The introduction of one piece flow from workstation 8 (dry Finishing) to 11 (gravelling) brought down the waiting time (WT), significantly. Overall waiting time is reduced from 24.84 to 16.75 days, i.e. an improvement of

8.09 days (32.56%) which is very significant. This has been achieved mainly by implementing following kaizens:

- After PED, the pugs were kept in the pug yard for some time with covering of polythene. This step was removed and pugs were directly transferred to the dryer after shaping (waiting time reduced by 0.67 days, WIP to zero).
- Single piece flow was created among workstations 8-9-10-11 by scheduled leveling. The processing time of dry-finishing (workstation 8) was 206 s and for other three stations it was low (varying between 102 and 106 s). One more station of dry-finishing was installed to make one piece flow. No new worker was assigned at new workstation 8, but two persons who were working for batch shifting in previous layout were used to do the processing at new dry-finishing workstation. This helped in better handling of material, reduction in motion time, reduction in waiting time and drastic reduction in floor inventory (WIP reduced, waiting time reduced, handling defects reduced).
- The kiln department was facing shortage of space. Efforts were made to reduce the inventory on the shop floor. It was observed that inventory was high in the post-kiln operations. Previously XXX was holding 7 days inventory in the warehouse because of poor communication and just-in-case tendency. Electronic information flow was started for the suppliers, which reduced the inventory to 6 days. It also helped in reducing order quantity and inventory at raw material stores in the company. Inventory level of raw material in store reduced to 2 months from 4 months. Inventory was reduced in pre and post kiln operations (Reduction in WIP along with reduction in WT after workstations 16, 18 and 21). Overall WIP is reduced by 36% starting from ball milling to packaging.

3.4.3 Space Creation

Lean initiatives, particularly 5S implementation, allowed a better use of the physical space and machinery and created 719 m^2 space after the case study.

3.4.4 Breakdown Reduction

The gain in break down time in shaping, C&G and testing processes/workstations after autonomous maintenance implementation is shown in Table 3.1.

Table 3.1 Average breakdown in the shaping workstation

S. No.	Name of workstation	Monthly average breakdown before autonomous maintenance (h)	Monthly average breakdown maintenance (h)
1	Shaping	6.14	3.52
2	C&G	7.60	3.90
3	Testing	8.88	3.88

3.4.5 Defect Reduction

Action/improvement initiatives were taken to reduce the defect/rejection rate. It was observed that the major defects/green rejections are in the category of Shed Bend, Shed Cut, Round Crack, Damaged Pug, Shaping, and handling rejections. The improvements of these initiatives are shown in Figs. 3.3, 3.4 and 3.5.

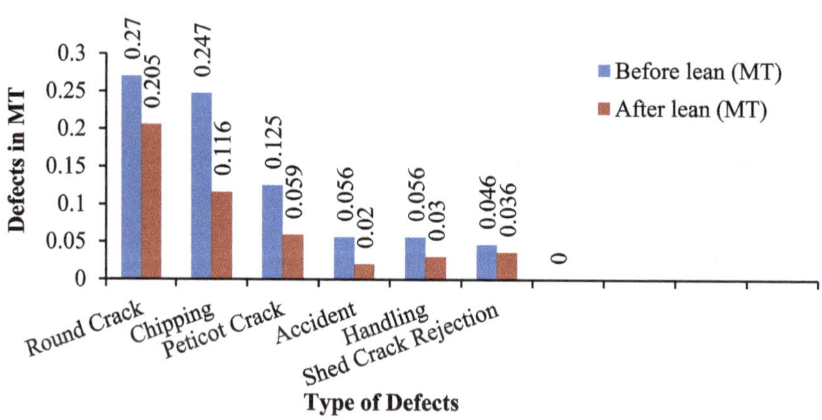

Fig. 3.3 Defect rate comparison at kiln workstation

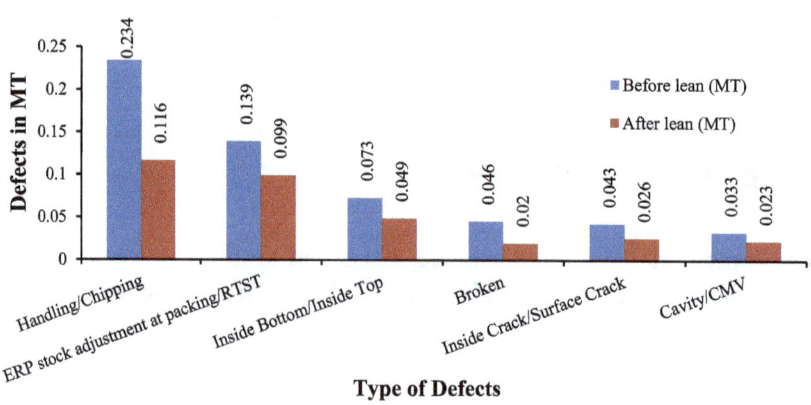

Fig. 3.4 Defect rate comparison in post kiln processes

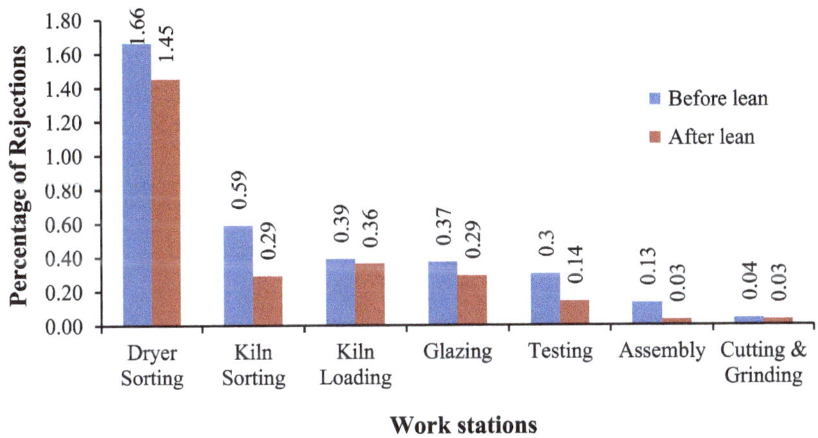

Fig. 3.5 Material handling rejections

3.4.6 Kiln Rejection

(See Fig. 3.3).

3.4.7 Post Kiln Rejections

(See Fig. 3.4).

3.4.8 Material Handling Rejection

Before lean manufacturing implementation, overall monthly post-kiln rejections was 3.258 MT and after LM implementation, it is 1.938 MT, i.e. an improvement of 40.51%. These calculations are based on monthly average defects.

3.4.9 Manpower Reduction

One-piece flow in dry finishing, glazing, dipping and gravelling reduced manpower used for transfer of products. The pug machine operator was engaged in starting/stopping the belt after extrusion and in marking the pugs after the rack is completely loaded. He was effectively working for 10% of the time; assigned another machine. Present conveyor system was replaced with a hydraulic operated lift system by which one operator was reduced. Also, kaizens at testing, pugging and kiln resulted in reduction of manpower. The total reduction in manpower is 6.85% (68 from 73).

3.5 Conclusions

This paper has shown the application of lean manufacturing in a semi-process industry. The organization has become flexible by eliminating wastes at various processes/workstations and responding to fluctuating customer demands quickly and efficiently. The various quantitative benefits of lean manufacturing implementation for the case organization are as:

- Waiting time reduction by 32%
- Processing time reduction by 191.56 min. (1.02%)
- Reduction in inventory/WIP by 36%
- Defects reduction (average) by 40.51%
- Space creation for further use is 719 m^2
- Reduction in manpower by 6.85%
- Cost saving of US $ 0.61 million in one financial year.

Qualitative benefits have also been observed in term of skill up gradation, team work, multi-skilling and improved morale of the employees.

References

Bhamu J, Sangwan KS (2014) Lean manufacturing: Literature review and research issues. Int J Oper Prod Manag 34(7):876–940

Bhamu J, Shailendra Kumar JV, Sangwan KS (2012) Productivity and quality improvement through value stream mapping: a case study of Indian automotive industry. Int J Prod Qual Manag 10(3):288–306

Carvalho CP, Carvalho DS, Silva MB (2018) Value stream mapping as a lean manufacturing tool: a new account approach for cost saving in a textile company. Int J Prod Manag Eng 7(1), (2018)

Jimenez E, Teieda A, Perez M, Martinez E (2011) Applicability of lean production with VSM to the Rioja wine sector||. Int J Prod Res 50(7):1890–1904

Narapinji P (2016) Waste reduction in manufacturing process: a case study of ceramic factories in Thailand 5:126–132

Rother M, Shook J (1999) Learning to see: value stream mapping to add value and eliminate MUDA. The Lean Enterprise Institute, Brookline, MA

Sangwa NR, Sangwan KS (2018) Leanness assessment of organizational performance: a systematic literature review. J Manuf Technol Manag 25(6):1746–1761

Yuvamitra K, Lee J, Dong K (2017) Value stream mapping of rope manufacturing: a case study. Int J Manuf Eng 1–11 (2017)

4

3D Printing and Decentralized Manufacturing

Lennart Büth, Max Juraschek, Sebastian Thiede, and Christoph Herrmann

Abstract As additive manufacturing technologies become widely available as desktop 3D printers, opportunities for decentralized manufacturing arise. The goal of the 3-CYCLE project is to produce plastic filament for 3D printing by recycling waste plastics and thus moving towards sustainable decentralized manufacturing. Leading to the question which product should be produced within the paradigm of sustainable distributed manufacturing using 3D printing with locally recycled filament. Therefore, this work proposes a framework to rate potential products regarding fabrication with recycled filament. A hierarchical criteria-system is developed covering important properties and constraints linked to the 3-CYCLE process and its characteristics. A weighting process to assign different importance levels to the proposed criteria-system by experts is created. Subsequently, a decision-making process is developed to rank in accordance of the assessed alternatives to the criteria-system. The multi-attribute process includes decision-making techniques (TOPSIS, WSM and WPM), an aggregation step and a verification approach. In a case study, ten products are compared with the developed approach. Resulting into four different rankings, differentiating between India, Germany and the inclusion of educational aspects.

Keywords 3D printing · Multi-criteria decision making · Recycled filament

4.1 Introduction

Plastic has become an essential part of daily life. The global production has increased by 500% over the last 30 years and will continue to grow to 850 million tons by 2050 (Shen et al. 2009). This trend is resulting in increasing environmental issues as for instance land (Rees 1980) and water (Derraik 2002) pollution rises. The project

L. Büth (✉) · M. Juraschek · S. Thiede · C. Herrmann
Chair of Sustainable Manufacturing and Life Cycle Engineering, Institute of Machine Tools and Production Technology (IWF), Technische Universität Braunschweig, Langer Kamp 19b, 38106 Braunschweig, Germany
e-mail: l.bueth@tu-braunschweig.de

3-CYCLE (the project name refers to the goal to recycle plastic material with 3D-printing) was established between the universities TU Braunschweig and Birla Institute of Technology and Science (BITS) Pilani to address the presented problems by utilizing the current open source development of fused deposition modelling (FDM). The Institute of Machine Tools and Production Technology of the TU Braunschweig designed an integrated plastic recycling process for the FDM 3D printing technology. This process is researched under the following constraints, which also serve as input for this work: The process should be resilient, be easy to perform and be transferable to an economical usage in a small, distributed scale. The deployment of this process in different settings and countries should have economic value and raise awareness of recycling possibilities and the recycling necessity of plastics. In cooperation with the Indian university BITS Pilani, this process is in development. Furthermore, the outcome should also be used in educational purposes as part of lectures and/or student projects (Juraschek et al. 2017).

This work addresses the later stage of the 3-CYCLE project. After the technical and the organizational processes were designed, the most effective utilization of this project has to be determined. Therefore, this work develops a methodological approach to rate potential outcomes of the introduced 3-CYCLE process. Starting with the produced filament as input material, the process should take a variety of criteria and boundary conditions into consideration. As a special requirement the work tries to include regional characteristics from India and Germany, as well as educational value of product ideas. With the result is a multi-criteria decision making (MCDM) methodology that is able to compare different ideas for this specified filament usage.

4.2 Criteria-System Definition

The requirements for the criteria-system as background for the rating methodology are defined by the goals of the 3-CYCLE project as stated in Chap. 1. To ensure a structured decision making process and to ensure the applicability of the criteria-system, formal and structural requirements need to be fulfilled:

Structured in a Hierarchy. The criteria-system should be structured in a hierarchy. On the one hand, a hierarchical system provides a clear structure. On the other hand, with a hierarchical order, the subsequent weighting process can be accomplished in incremental steps. Without a hierarchical structure, the weighting process has to compare heterogeneous information, e.g. specified (second-tier) economically criteria to specified social criteria. Resulting in potentially flawed results. Such a hierarchical criteria-system is e.g. proposed by Mendoza and Macoun (1999).

Formulated as Verifiers. All selected criteria have to be formulated as verifiers and have to be measurable with available tools, within available capabilities and data. Whenever possible, objective verifiers should be preferred over subjectively rated verifiers.

Addressing Feasibility. An important requirement is the technical feasibility of the production process of FDM 3-D printing. Limitations and advantages should be considered in a criteria-system. Castillo et al. (2012) including the feasibility into the principle of human-centered design. Besides the feasibility, the viability and the desirability of a product should be addressed. Therefore, the feasibility is defined with the topics of design, technology and innovation, summarized with the question "What is technically possible?". The viability of a product is defined with the question "What can be financially viable?". And the desirability is defined as user context research, with the leading question "What do users desire?" (Kelley et al. 2009).

Addressing Sustainability. Fostering sustainable development, economic, environmental and social impacts of the recycling and production process as well as of the subsequent life cycle phases should be considered. Also included as criteria by (Savonen 2015) for sustainable product design for 3D printing.

Addressing Academia. Lastly, the criteria shall address the need to develop the project close to aca-demia and therefore the product should have an educational value inherent.

Based on the derived requirements a hierarchical criteria-system with three tiers is created as illustrated in Fig. 4.1. The principles are set at the highest tier and consist of human-centered design, sustainability and the conceptual-fit. While the first two principles are self-explanatory, the principle conceptual-fit is a place holder for giving context to the circumstances in which the system is applied. In this work it fulfills the educational context as stated in requirements (5). Following, the criteria tier is clustering the principles to further ensure a more objective weighting process, while consisting of criteria described in the requirements (3–5).

In the lower tier, indicators describe the criteria more specific. Each indicator is operationalized by verifiers for the later decision making process. All indicators/verifiers are stated within Table 4.1. The verifiers are designed heterogeneously, ranging from qualitative scales to measurements in degrees or distances. Qualitative scale ranges are selected due to the granularity of possible alternatives, aiming for linear coherences, a high value is desired. This property results in the requirements for the decision-making process to handle heterogeneously verifiers.

4.3 Methodological Approach for Multi-criteria Decision Making

The method developed aims at evaluating possible products with regard to the characteristics of the 3-CYCLE and the developed criteria-system. The approach should be able to weight the heterogeneous verifiers. The process is based on the approach for sustainable decision making presented in Wang et al. (2009). Adopting the procedure and the main elements: Criteria selection (Chap. 2), weighting, multi-criteria decision making and aggregation. The overall approach is summarized in Fig. 4.2.

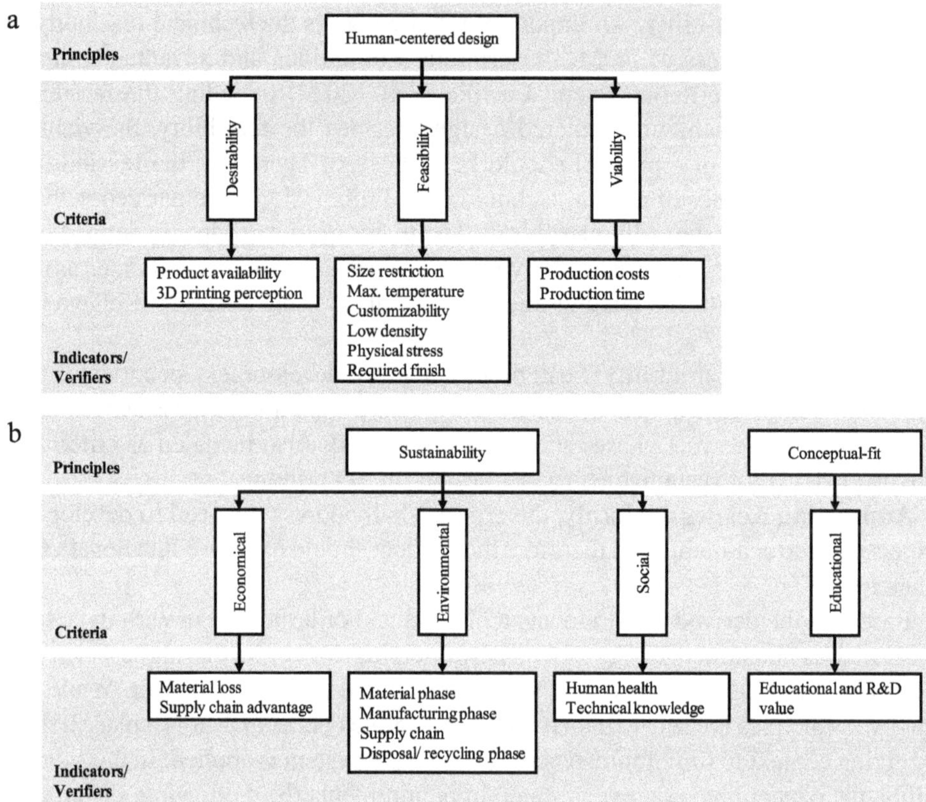

Fig. 4.1 Hierarchical criteria-system, consisting of three tiers: Principles, criteria and indicators/verifiers. **a** Principle Human-centered design, **b** principles Sustainability and Conceptual-fit

4.3.1 Weighting Process Using AHP and Delphi Iterations

The weighting process for the criteria is combining Analytic Hierarchy Processes (AHP) (Saaty 1988) as well as Delphi iterations (Wang et al. 2009) (see grey area Fig. 4.2). This approach enables the combination of weights created by multiple experts. Within an AHP the relation of two criteria rated by a pairwise comparison resulting in a weighting. Each individual expert starts with an AHP weighting process in seven steps:

1. Introduction of the topic
2. Introduction of the hierarchical criteria-system
3. Specific explanation of the criterion-indicators
4. Successive seven AHPs, indicator level
5. Specific explanation of the criteria of a principle
6. Successive three AHPs, criterion level
7. AHP, principle level.

Table 4.1 Overview of all indicators, verifiers and the respective measurement scales of all three principles, utilizing heterogeneous scales

Indicator	Verifier	Scale
Product availability on markets	Scale	0–24
Changed perceived value due 3DP Scale	Scale and score	0–3
Max. temperature	Temperature	°C
Customizability level	Scale	0–4
Low density advantage	Scale	0–2
Physical stress	Scale and score	0–12
Required finish	Decision	Yes/No
Production costs	Costs	Euro
Production time	Time	Minutes
Material loss	Malus matrix	1, 2
Supply chain advantage	Scale	0–3
Impact in raw material phase	CO_2 Eq.	Difference
Impact in manufacturing phase	CO_2 Eq.	Difference
Impact in recycling/deposal phase	CO_2 Eq.	Difference
Impact in supply chain	Scale	0–3
Human health, due to post-processing	Decision	Yes/No
Requirement technical knowledge	Scale	0–2
Educational and R&D value	Scale	0–2

After conducting initial AHPs with multiple experts, Delphi iterations follow. Therefore, the resulting weightings of all experts are aggregated to a merged weighting. Then the individual experts are exposed to the merged weighting. The individual experts can revisit their weighting with regard to the merged weighting, resulting in a second iteration of individual AHP results. In a final step these results are aggregated to the final criteria-weighting.

4.3.2 Multi-attribute Decision Making

In the next step, the product alternatives are rated according to the criteria-system applying the verifiers. After values are assigned, the data is normalized and thereby prepared for the MCDM techniques.

Many different MCDM methods are available in the literature. The subset of MCDM suitable for the decision problem of this work is called multi-attribute decision making (MADM), as the criteria are attributes of the set products. Different

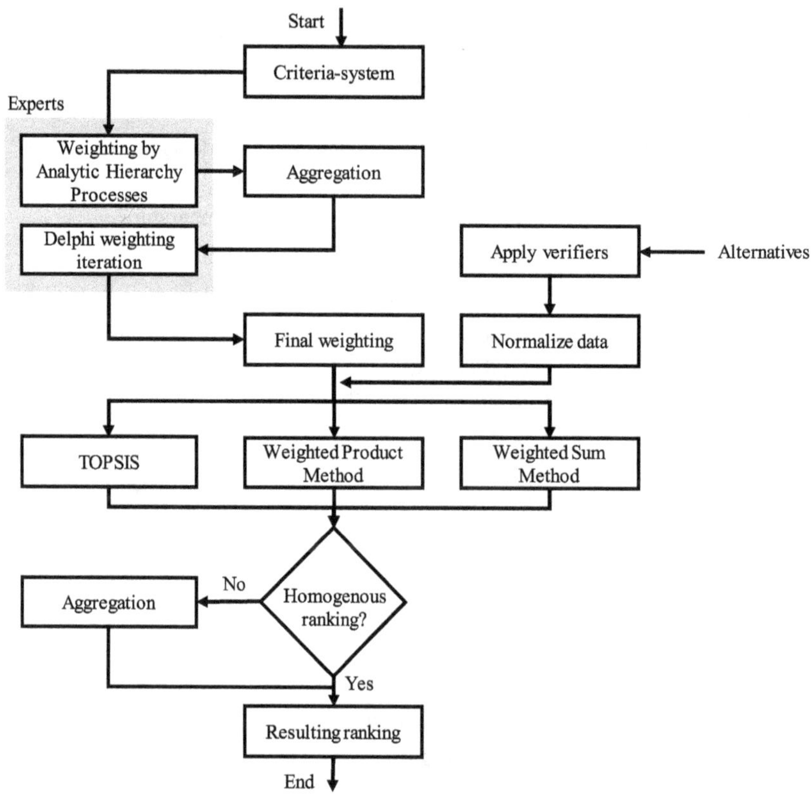

Fig. 4.2 Overview of the developed methodological approach for MADM using the developed criteria-system

MADM approaches are used in literature and no clear advantages on one or the other can be found. Therefore, a literature review was conducted on similar decision problems. Appling the three most used MADM methods for product selection: Technique for Order Preference by Similarity to Ideal Solution (TOPSIS), Weighted Product Method (WPM) and Weighted Sum Method (WSM). These techniques are executed in the next step, with the input of the normalized data and the weights. The resulting rankings are then compared and if necessary aggregated.

To ensure all crucial parts are considered in the decision-making process, the developed process is compared to the generalized approach for sustainable decision making from WANG et al. (2009). All steps before the beginning of the weighting procedure are covered. The loop starting with "Checking the concordance of criteria values", seem to be different, but is similar to the Delphi iteration process. Instead of increasing the number of experts, the experts can reevaluate their weights. Proceeding, WANG et al. determine objective weights, and check these for concordance/integrate them. As previously discussed, a decision between subjective and objective weighting (e.g. equal weights) was done. Therefore, this loop was not adopted in the developed process. The aggregation loop of WANG et al. matches the present aggregation procedure. To conclude, all crucial elements were considered, only the objective weighting loop is missing out. If necessary such a loop could be integrated, when improving the approach.

4.4 Case Study: What to Use Recycled Plastic Filament for?

In a case study ten different 3D printable products were exemplary compared using the developed approach summarized in Fig. 4.2. The following 3D printable products were ranked: Calipers, a wardrobe wall adapter, a hand prosthesis, a phone case, a bracelet, a soap case, a robot arm kinematic, a tape gun, a whistle and an ear bud holder. The approach was applied for four different rakings: Usage in India and Germany with as well as without the principle conceptual-fit. The results are visualized in Fig. 4.3, "1" being the best ranked and "10" the worst.

Exemplary the rating of the product alternative "Bracelet" for the criterion feasibility is stated: Build size restriction is the first feasibility verifier, the bracelet fits on all printers building platform and is therefore assigned with the value "2". The temperature limits of PET are not exceeded in the using phase of a bracelet; thus the verifier temperature is rated with "No". The customizability level scale is rated with "2", as the outer appearance of the bracelet is its key feature and potentially total customizable. Regarding the verifier low density advantage, the value "0" is assigned. A bracelet has a relatively thin design, and can therefore not utilize the low density advantage of 3D printing. The verifier physical stress is rated with a "9". In the subcategories the stress categories tension, flexural, fatigue and wear stress, got rated with the maximum value "2". A bracelet, as an everyday wearing device, is strongly exposed to these kinds of physical stress. The category compression stress

Fig. 4.3 Aggregated MADM results, comparing ten products in four different ratings

		Rating			
		India	Germany	India incl. Conceptual-fit	Germany incl. Conceptual-fit
Alternatives	Calipers	5	3	3	2
	Wardrobe	5	6	10	10
	Prothesis	2	10	1	8
	Phone Case	7	4	8	7
	Bracelet	1	1	2	1
	Soap Case	3	4	4	6
	Kinematic	10	9	4	3
	Tape Gun	9	7	7	3
	Whistle	3	2	6	3
	Bud holder	8	7	9	9

is rated with "1", as compression stress can occur during handling process but only limited in the intended use case. The category impact stress was assigned with the value "2", as no physical impact should occur in an intended using scenario. The last feasibility verifier required finish was rated with "No". No finish is mandatory for the bracelet to fulfill its application.

The final results over all normalized criteria of the aggregated MADM methods are stated in Fig. 4.3. Some alternatives are homogenous rated between all four iterations (e.g. the alternatives ear bud holder and bracelet). Some alternatives clearly state the difference between the regions and between the inclusion and exclusion of the conceptual-fit principle.

The alternative wardrobe is clearly showing the effect of the included conceptual-fit. Due to the relatively bad rating in this high weighted verifier, the wardrobe is ranked 10 for both regions. In contrast the robot arm kinematic is relatively better ranked due to the high impact of good ratings in the added principle. The alternative hand prosthesis is indicating the differences between both regions, most likely due to the impact of two verifiers measuring impact and advantage of the supply chain. Additionally, the verifiers production costs (e.g. differentiating energy costs) and product availability are contributing to the differences between India and Germany.

4.5 Discussion and Conclusions

A decision making approach for potential 3-CYCLE products was developed and applied in a case study. The weighting process was successfully conducted. The second weighting iteration yielded only minor changes to the first round results, this might be reasoned due to the relatively homogenous expert group. The feedback of the experts regarding the hierarchical system design was positive. Sometimes it was difficult for the experts to remember the definitions of the introduced criteria/indicators while conducting the AHPs. Assigning values to the alternative/indicator combination was unproblematic. Only the price of the prosthesis had to be assumed. Some of the introduced scales were easy to rate (e.g. Yes/No verifiers), some scales are relying on assumptions. The designed MADM process is mainly relying on subjective opinions, as is the criteria-system. Therefore, an improvement regarding this coherence may be necessary in future. Although the case study proves that the developed approach fulfills all requirements of the problem. The case study provided an initial insight in the application of the developed method, the results give a first orientation for the 3-CYCLE project.

References

Shen L, Haufe J, Patel M (2009) Product overview and market projection of emerging bio-based plastics PRO-BIP 2009. Report for European polysaccharide network of excellence (EPNOE) and European bioplastics 243

Rees J (1980) The fate of carbon compounds in the landfill disposal of organic matter. J Chem Technol Biotechnol. https://doi.org/10.1002/jctb.503300121

Derraik JG (2002) The pollution of the marine environment by plastic debris: a review. Mar Pollut Bull 44(9):842–852. https://doi.org/10.1016/S0025-326X(02)00220-5

Juraschek M, Cerdas F, Posselt G, Herrmann C (2017) Experiencing closed loop manufacturing in a learning environment. Procedia Manuf 9:57–64. https://doi.org/10.1016/j.promfg.2017.04.046

Mendoza G, Macoun P (1999) Guidelines for applying multi-criteria analysis to the assessment of criteria and indicators. The criteria & indicators toolbox series, 60, center for international forestry research. https://doi.org/10.17528/cifor/000769

Castillo LG, Diehl JC, Brezet JC (2012) Design considerations for base of the pyramid (BoP) projects. In: Proceedings of the cumulus helsinki 2012 conference, pp 1–15. https://doi.org/10.1145/1865245.1865255

Kelley D, Moggridge B, Nuttall M (2009) Design kit: the field guide to human-centered design

Savonen BL (2015) Criteria for sustainable product design with 3D printing in the developing world, Technical report Michigan Technological University

Wang JJ, Jing YY, Zhang CF, Zhao JH (2009) Review on multi-criteria decision analysis aid in sustainable energy decision-making. Renew Sustain Energy Rev 13(9):2263–2278. https://doi.org/10.1016/j.rser.2009.06.021

Saaty TL (1988) What is the analytic hierarchy process? mathematical models for decision support. Springer, Heidelberg, pp 109–121. https://doi.org/10.1007/978-3-642-83555-1_5

5

Solar Thermal Devices: Power Generation and Power Cycle

Manoj Kumar Soni and Aditya Soni

Abstract Use of solar thermal devices has proved a viable option for power generation in many countries that have abundant solar radiation throughout the year. Power Tower and Parabolic Trough collectors are the most common and commercially proven technologies in such solar thermal power plants. This paper presents the design and simulation of 1 MWe solar thermal power plant in which the steam is generated by Australian Nation University's (ANU) 520 m^2 solar parabolic dish concentrators. The plant also powers a vapour absorption-chilling unit that uses steam extracted from the turbine to produce a cooling effect. This paper discusses the design of the power cycle, and through the simulation, estimates the annual power generation from the power block and tonnes of refrigeration produced by vapour absorption chiller. Apart from the design of the power cycle, the layout of dishes is also designed for maximum steam generation from the solar field. The waste heat from the vapour absorption system may be used for low-grade heating applications.

Keywords Solar thermal power plant · Solar refrigeration · Tri-generation

5.1 Introduction

Energy is the major requirement in the development of any country. At the global level, the demand for primary energy consumption is met mainly by fossil fuels, which are the major contributor to global warming. In order to mitigate this, renewable energy sources which are inexhaustible is the key solution. Amongst various renewable energy sources solar and wind have the major stake in world energy

M. K. Soni (✉)
Mechanical Engineering Department, Birla Institute of Technology and Science, Pilani-Campus, Pilani 333031, RJ, India
e-mail: mssoni@pilani.bits-pilani.ac.in

A. Soni
Sunrise CSP India Pvt. Ltd, Muni Seva Ashram, Goraj, Vadodara, India
e-mail: aditya.soni@sunrisecsp.com

supply through renewables. As per ETP, 2015, these two sources have the potential to reduce 22% of annual emissions in 2050. In solar energy especially the solar thermal in general and concentrating solar power (CSP) in particular has the potential to provide 7 and 25% of the world electricity by 2030 and 2050 respectively (Ummadisingu and Soni 2011; Sunil 2017; Gakkhar et al. 2016; Soni and Gakkhar 2014; Sunil and Soni 2019).

India with a strong economic growth rate is one of the fastest developing economies in the world. In order to sustain and accelerate this growth rate further, the demand for energy will grow simultaneously. The supply of this required energy should be secured, sustainable, affordable and environmental friendly also. For this solar energy plays a very important role as India is blessed with 250–300 clear and sunny days. Most part of the country receiving 4–7 kWh/m^2 of solar radiation incident over India's land area, which is equal to 5 trillion MWh per year energy (Sunil 2017; JNNSM 2009; Energy Technology Perspectives 2015).

5.1.1 Solar Power Generation

Solar energy is utilised to generate power in two different manners. First is the use of Photovoltaic effect of light-absorbing materials, for example, silicone, to directly generate electricity. The second way to utilise solar energy is to use the heat aspect of the sun's radiation as a thermal source of energy. In conventional thermal power plants, coal is used to generate steam. Instead of burning coal, solar thermal energy can be used to generate steam, reducing the carbon emissions from the power plant drastically.

Major components of a standard thermal power plant are turbine, condenser, pumps and boiler. In solar thermal power plants, the steam generation is done by solar concentrators instead of boilers (Nayak et al. 2015). Auxiliary boilers are usually present to provide backup thermal power for non-sunshine hours.

5.1.2 Solar Space Cooling

Producing cooling effect from solar thermal energy is possible through vapour absorption refrigeration system (VARs). The VARs in common use are (1) aqua-ammonia system with ammonia as refrigerant and water as absorbent, (2) water - lithium bromide (Li Br) system where LiBr is an absorbent and water as a refrigerant. Major components of such units include components like absorber, small pump, generator, condenser, expansion device and evaporator. In VARs, the refrigerant gets absorbed in the absorber, the strong solution is then pumped to generator raising its pressure. In generator, the heat is supplied, which may through solar energy or any waste heat available. Due to this heat addition, the solution of absorbent and refrigerant is heated. As it is heated over the boiling point of the refrigerant, the refrigerant gets separated from the absorbent. The refrigerant vapours are then condensed in the condenser and the condensate is then throttled to the evaporator pressure and then gets evaporated in the evaporator producing cooling effect. The refrigerant produces a cooling effect in the evaporator and releases the heat to the atmosphere via the condenser. Le Lostec et al. (2013) have done a numerical simulation of aqua ammonia vapour absorption chiller and validated the results with the experimental data.

5.1.3 Cogeneration

Cogeneration, also called *Combined Heat and Power* is a systematic way to draw some minor fraction of steam from the turbine for an application that may require heat (Sahoo 2018). In this paper, the steam required by the VARs chiller is taken out of the turbine at an intermediate pressure.

Colonna and Gabrielli (2003) analysed an industrial tri-generation system producing electrical power, heat and cooling. They analysed VARs with waste heat recovery from internal combustion engines in the form of pressurized hot water, which is then supplied to VARs for the cases of Italy and Netherlands. Behnam et al. (2018) analysed a small-scale tri-generation system with a geothermal energy source to produce power using organic Rankine cycle-based power plant, LiBr based VARs and heat for desalination.

5.2 Design of Power Plant

5.2.1 Site Selection

The site under consideration is Muni Seva Ashram at Goraj, Vadodara. The site is adjacent to the existing solar field of 100 Scheffler dishes. Dev river runs nearby providing water to the ashram. Water from the same river is considered to be used for running the power plant and its cogeneration elements after treatment. The utility building of the ashram which houses a boiler, backup diesel gen-sets and centralized refrigeration system is also adjacent to the site. The building has enough space to house the turbine, condenser and auxiliary boiler unit. The site has ample solar radiation for about 300 sunny days. The monthly average direct normal irradiance (DNI) and monthly average temperature for the site have been shown in Fig. 5.1.

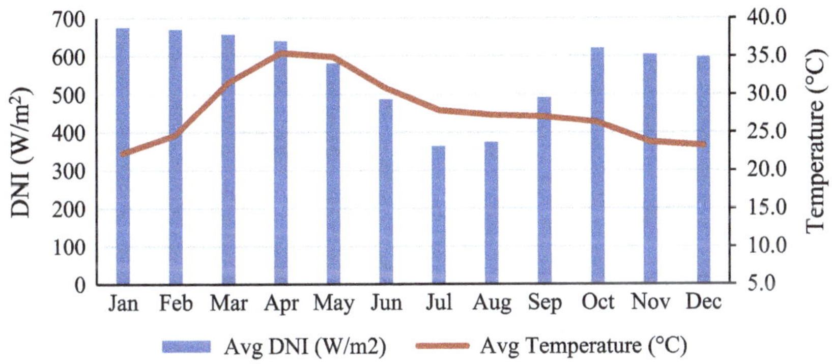

Fig. 5.1 Monthly-averaged DNI and temperature data

5.2.2 Solar Field

The solar concentrators used in the system are solar generator 4 (SG4) BigDish, developed by ANU (Lovegrove et al. 2011). These are 520 m^2 parabolic dishes with dual-axis tracking fitted with a high-efficiency receiver (Pye et al. 2016). Analysing the solar radiation data, it is found that the site has the potential to provide optimum solar energy from 8 to 16th h of a typical day. As the NREL data counts the hour between 00:00 and 01:00 h as the zeroth hour, operation from 8 to 16th h will be 08:00–17:00 h, 9 h in total.

The layout of the dishes should be such that the adjacent dishes do not obstruct and cast a shadow when the sun is on the horizon. For the location Elevation (α) and Azimuth (γ) angles were calculated. It was found that between 8th to 16th h for each day throughout the year, the smallest value of α is 14.10°. This was in the morning, on the 8th h of 11th January. Corresponding γ angle was found to be 58.74°. The parabolic dishes were arranged with this orientation in mind. At this orientation, there should not be any shadow being cast on the adjacent dish. All other values of α and γ will be so that they will not cast any shadow. For 1 MW output from the system, the solar field designed consists of 14 SG4 solar dishes of 520 m^2 size will be required considering designed DNI of 700 W/m^2. Figure 5.2 shows the placements of the proposed solar field near Muni Seva Ashram.

5.2.3 Design of Power Cycle

A power cycle is based on regenerative Rankine cycle with condensing turbine. Turbine has two bleeds, one goes to VARs chiller and other to the deaerator. The bleed to VARs depends on the cooling load requirements. Lower the demand more will be the steam available for the turbine to generate power and condense. Figure 5.3 shows the block diagram of the power cycle with VARs machine (VAM) and deaerator. The steam from the solar field enters the turbine at 65 bar(a) and 360 °C. The isentropic

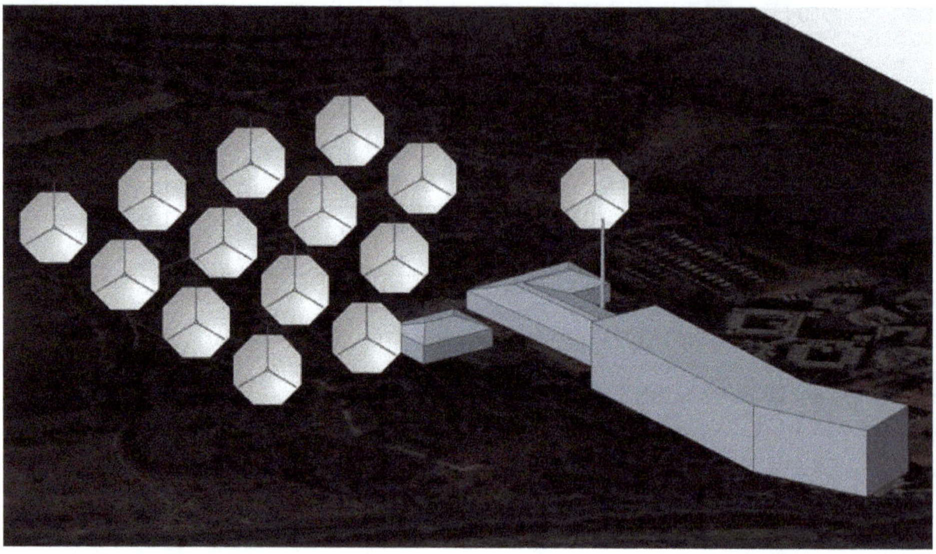

Fig. 5.2 Placement of the proposed solar field near Muni Seva Ashram

Fig. 5.3 Block diagram of power cycle with VAM

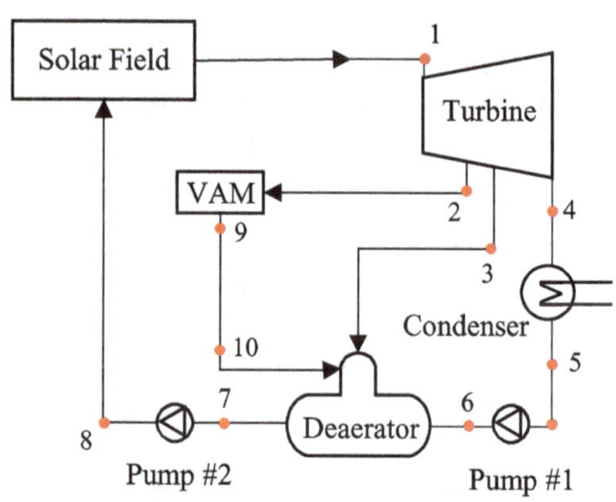

efficiency of the turbine is considered as 60%. Though the proposed system is capable of producing steam at pressure and temperature of the order of 160 bar and 500 °C, the turbine inlet pressure and temperature were selected based on 1 MW turbine available in the market. The first extraction of the steam at 8.5 bar(a) and 174.5 °C with a flow rate of 360 kg/h is supplied to VAM, and second extraction at 4.76 bar(a) is given to deaerator, remaining steam after expansion in turbine comes out at 0.2 bar(a) and goes to condenser. The condensate from the condenser is then pumped to the deaerator where it meets the condensate from VAM and extraction from the turbine. The condensate from the deaerator is then pumped to the boiler pressure and sent back to the solar field.

5.2.4 *Vapour Absorption Chiller Calculations*

The existing requirement for cooling to be met at the ashram is about 80 tonnes of refrigeration (TR) so a VAM chiller to provide 80 TR for space cooling is selected. Looking at the daily temperature profile of the location, the maximum temperature was found to be 45 °C. The required tonnage of cooling is reduced at lower temperatures, assuming that all the other factors that contribute as heat loads do not vary. The VARs selected is a Double Effect VAM Chiller using LiBr-Water solution as a refrigerant with an operating pressure of 8.5 bar(a) with the steam flow rate of 360 kg/h at full load.

5.3 Results and Discussion

The proposed system has been simulated using MATLAB and the results of the simulation are presented in Figs. 5.4, 5.5 and 5.6. Figure 5.4 shows the monthly average refrigeration supplied, and the results represent a total of 15870 TR, an average of 43.47 TR each day of the year, with a peak in March, April and May

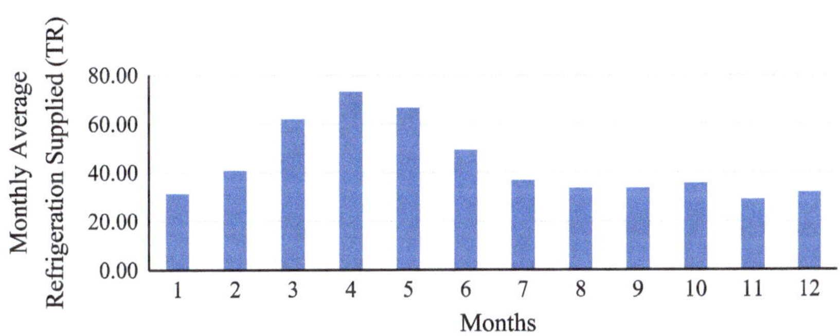

Fig. 5.4 Monthly average refrigeration supplied (TR)

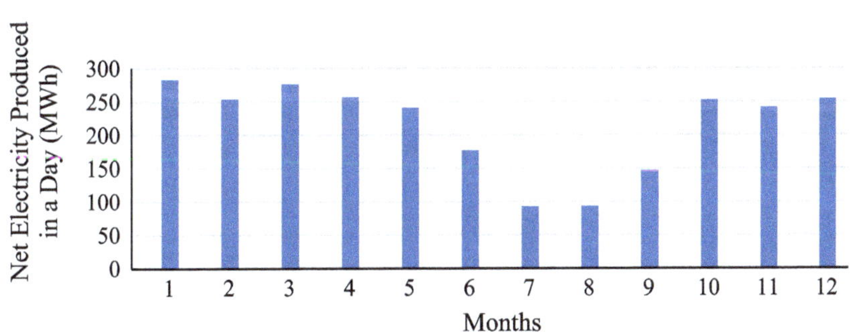

Fig. 5.5 Monthly average electricity generation (MWh)

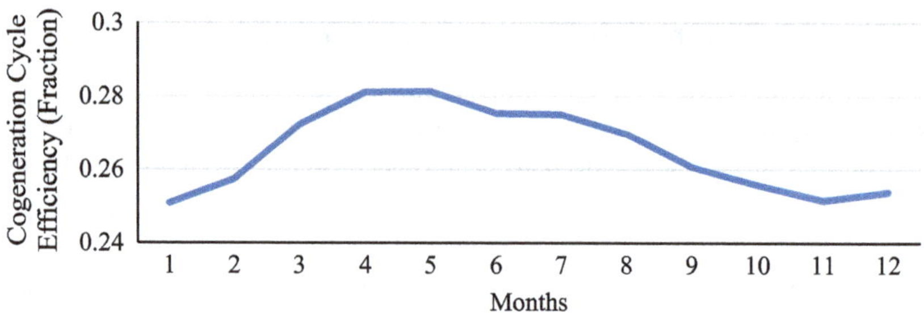

Fig. 5.6 Cogeneration cycle efficiency (Fraction)

when it is very much in demand. Figure 5.5 represents the monthly average electricity generation with a total of 2566 MWh of electricity generation throughout the year with about 250 MWh per month for almost eight months. Figure 5.5 represents cogeneration cycle efficiency, the results show that the system as a whole is expected to work within the range of 25–28% efficiency, which is a very good performance for the 1 MW scale power systems. The waste heat from the absorber and condenser of VAM may be utilized for low-temperature heating applications like the supply of hot water, or it may also be used for distillation application. Such a combined system will give power, cooling as well as heat and hence tri-generation system.

5.4 Conclusion

This paper presents the design and simulation of 1 MWe solar thermal power plant-based tri-generation system. This system is designed for the location of Muni Seva Ashram, at Goraj, Vadodara in Gujarat, India. The data analysis shows that the proposed site has sufficient DNI to run such a system.

The solar-based steam generator using ANU's 520 m^2 solar parabolic dish concentrators is used in the proposed tri-generation system to supply steam at 65 bar(a) and 360 °C producing 2566 MWh of electrical power generation throughout the year. The steam extracted from turbine powers a vapour absorption chilling unit to produce 15870 tons of refrigeration throughout the year with a peak in March, April and May when it is very much in demand. The proposed system is expected to work within a range of 25–28% of efficiency. The waste heat from the vapour absorption system may be used for low-grade heating applications.

Apart from the design of the power cycle, the layout of dishes is also designed for maximum steam generation from the solar field.

References

Behnam P, Arefi A, Shafii MB (2018) Exergetic and thermoeconomic analysis of a trigeneration system producing electricity, hot water, and fresh water driven by low-temperature geothermal sources. Energy Convers Manag 57:266–276

Colonna P, Gabrielli S (2003) Industrial trigeneration using ammonia-water absorption refrigeration systems (AAR). Appl Therm Eng 23(4):381–396

Energy Technology Perspectives (2015) International energy agency. https://webstore.iea.org/download/summary/542?fileName=Chinese-ETP-2015-ES.pdf. Accessed 03 July 2019

Gakkhar N, Soni MS, Jakhar S (2016) Second law thermodynamic study of solar assisted distillation system: a review. Renew Sustain Energy Rev 56:519–535. https://doi.org/10.1016/j.rser.2015.11.076

JNNSM. (2009) Mission document, ministry of new and renewable energy government of India. Jawaharlal Nehru national solar mission. https://mnre.gov.in/sites/default/files/uploads/mission_document_JNNSM.pdf. Accessed 03 July 2019

Le Lostec B, Galanis N, Millette J (2013) Simulation of an ammonia-water absorption chiller. Renew Energy 60:269–283

Lovegrove K, Burgess G, Pye J (2011) A new 500 m^2 paraboloidal dish solar concentrator. Sol Energy 85(4):620–626

Nayak JK, Kedare SB, Banerjee R, Bandyopadhyay S, Desai NB, Paul S, Kapila A (2015) A 1 MW national solar thermal research cum demonstration facility at gwalpahari. Haryana, India, Curr Sci 109(8):1445–1457

Pye J, Hughes G, Abbasi E, Asselineau C.-A, Burgess G, Coventry J, Logie W, Venn F, Zapata J (2016) Development of a higher-efficiency tubular cavity receiver for direct steam generation on a dish concentrator. In: AIP conference proceedings

Sahoo U (2018) A polygeneration process concept for hybrid solar and biomass power plant: simulation, modelling, and optimization. Wiley, India

Soni MS, Gakkhar N (2014) Techno-economic parametric assessment of solar power in India: a survey. Renew Sustain Energy Rev 40:326–334. https://doi.org/10.1016/j.rser.2014.07.175

Sunil SMS (2017) Thermo-economic analysis of a solar-coal hybrid power plant. In: Proceedings of the 24th national and 2nd international ISHMT-ASTFE heat and mass transfer conference (IHMTC-2017). BITS Pilani, Hyderabad, India. https://doi.org/10.1615/ihmtc-2017.1940

Sunil SMS (2017) Integration of solar thermal energy in existing subcritical coal-fired power plant for fuel saving. In: ICAER 2017, 6th international conference on advances in energy research 2017. IIT Bombay, Mumbai, India

Sunil B, Soni MS (2019) Comparative performance assessment of a hybrid solar-coal power plant. Process Integr Optim Sustain 3(2):227–235. https://doi.org/10.1007/s41660-018-0066-x

Ummadisingu A, Soni MS (2011) Concentrating solar power—technology, potential and policy in India. Renew Sustain Energy Rev 15(9):5169–5175. https://doi.org/10.1016/j.rser.2011.07.040

6

Promoting Leanness: Application of Continuous Kaizen

Narpat Ram Sangwa and Kuldip Singh Sangwan

Abstract This study aims at introducing the concept of continuous kaizen to improve the leanness by enhancing productivity, line balancing and line efficiency of an automotive component assembly line. The present study shows the implementation of continuous kaizen at an assembly line using case study methodology. The proposed continuous kaizen implementation methodology illustrates how the top management specified goals/targets for the overall improvement of the line are achieved. The quality control techniques of Gemba walk, 3M (muda, mura, muri) analysis and ECRS (eliminate, combine, reduce, or eliminate) study are used for the micro analysis of the activities. It is found that the continuous kaizen can be effectively implemented in assembly lines to improve leanness by enhancing productivity, line balancing and line efficiency. It has been shown through the case study that the new concept of continuous kaizen decreased cycle time from 80 to 75 s, increased productivity by 6.7%, and line efficiency by 2.9%. The line balancing is also improved by decreasing the cycle time variation (standard deviation) from 4 to 2.84 σ.

Keywords Continuous kaizen (CK) · Leanness · Assembly line

6.1 Introduction

The organizations are implementing continuous improvement programs based on the philosophies of lean manufacturing, six sigma, total quality management (TQM), etc. to enhance customer satisfaction (Yang et al. 2016). The continuous improvement is a culture of consistent enrichment aiming at the identification and elimination of lean waste from the processes of an organization (Bhuiyan and Baghel 2005).

N. R. Sangwa
Department of Mechanical Engineering, Manda Institute of Technology, Bikaner, India

K. S. Sangwan (✉)
Department of Mechanical Engineering, Birla Institute of Technology and Science, Pilani Campus, Pilani 333031, RJ, India
e-mail: kss@pilani.bits-pilani.ac.in

Continuous small and incremental kaizen activities over a period of time lead to large improvements. The kaizen is a kind of project having the potential for the improvement of worker involvement and the organizational performance at the same time (Farris et al. 2008).

Chung (2018) claimed that kaizen is not similar to "improvement" in its usual nous. Suárez-Barraza et al. (2011) identified two interpretations of kaizen: the western explanation of kaizen as "continuous improvement" and the Japanese interpretation of kaizen as improvement by involving everyone alike. In western countries, the continuous improvement is termed as the kaizen and seen as a corporate proficiency that is practised as a part of either TQM or various other innovation and improvement programs (Bessant 2003). Whereas in Japan, the kaizen is described as a philosophy of conducting improvement activities at the workplace by involving everyone alike (Imai 1986). Aoki (2008) highlights the need to comprehend not only the execution of kaizen activities but also the spirit of kaizen in more depth. Thus, present study proposes a new delineation of kaizen—'continuous kaizen (CK)'. The paper defines 'continuous kaizen' as continuous and comprehensive improvement for the completeness at the global or whole value chain level instead of just 'change for better' at local or single workstation, so as to imply the value of integration in kaizen activities. The 'continuous kaizen' focuses on three key aspects:

- Kaizen should be throughout the value chain—the kaizen should be small, incremental, continuous, and comprehensive improvement.
- Kaizen should involve everyone from everywhere—the multi-hierarchical cross-functional team should perform kaizen activities in a specified timeframe to achieve pre-defined goal(s).
- Kaizen should improve leanness—the kaizen should improve the leanness level of the organization by systematic identification and elimination of various lean wastes.

There is still scant research on the kaizen activities in automotive component organizations (tier one supplier) as compared to automotive organizations (Marin-Garcia et al. 2009) and the implementation of the kaizen is also lesser outside the Japan (Aoki 2008). The meagre research on kaizen at the automotive assembly lines is best described by the difficulties associated with the implementation of kaizen at assembly lines. Leanness improvement of an automotive assembly line is challenging to improve as the number of processes involves a large number of components and sub-assemblies to make the final product (Salzman 2002). Assembly processes are prone to errors due to their complexity, resulting in higher costs and longer cycle times. The assembly processes are difficult to map and examine as compared to production. This case study demonstrate the implementation of continuous kaizen at an auto component assembly line to improve leanness.

6.2 Literature Review

Kaizen focuses on problem identification and its root causes and provides the creative solutions (Vonk 2005). The kaizen activities are used for the value addition to products and/or services (Marin-Garcia et al. 2018). The kaizen implementation attracts many organizations since it provides several qualitative and quantitative benefits to the organizations. The qualitative benefits are often related to human resource such as improvement in worker skills and commitment (Marin-Garcia et al. 2009); self-esteem and motivation (Alukal and Manos 2006); staff participation, training, communication, teamwork, and greater job satisfaction (Alvarado-Ramírez et al. 2018; Suárez-Barraza and Ramis-Pujol 2010). The quantitative benefits are linked to the economic factors such as increased productivity, profit, and inventory turnover (Oropesa et al. 2016); reduced lead times, cost, defects, and number of stages in production processes (Ramadani and Gerguri 2011).

The numerous tools, techniques and methods for kaizen implementation exist in the literature (Marin-Garcia et al. 2018). Typically, kaizen tools and techniques are human-based and process-oriented, while kaizen itself is continuous, incremental, and hands-on in nature (Suárez-Barraza et al. 2011; Alvarado-Ramírez et al. 2018). Suárez-Barraza and Lingham (2008) also identified four dimensions of kaizen—office kaizen, Gemba-kaizen, kaizen blitz, and kaizen teian. Marin-Garcia et al. (2018) have identified eight different types of tools and techniques for the implementation of kaizen activities—quality circles, ad hoc groups, suggestion systems in permanent teams, kaizen blitz, improvement teams, self-regulated work teams, and kaizen event.

6.3 Research Methodology

The research methodology for continuous kaizen is shown in Fig. 6.1.

6.4 Case Study

The present study illustrates a case of Indian automotive component assembly line, where management is worried about the challenges of lower productivity due to higher cycle time.

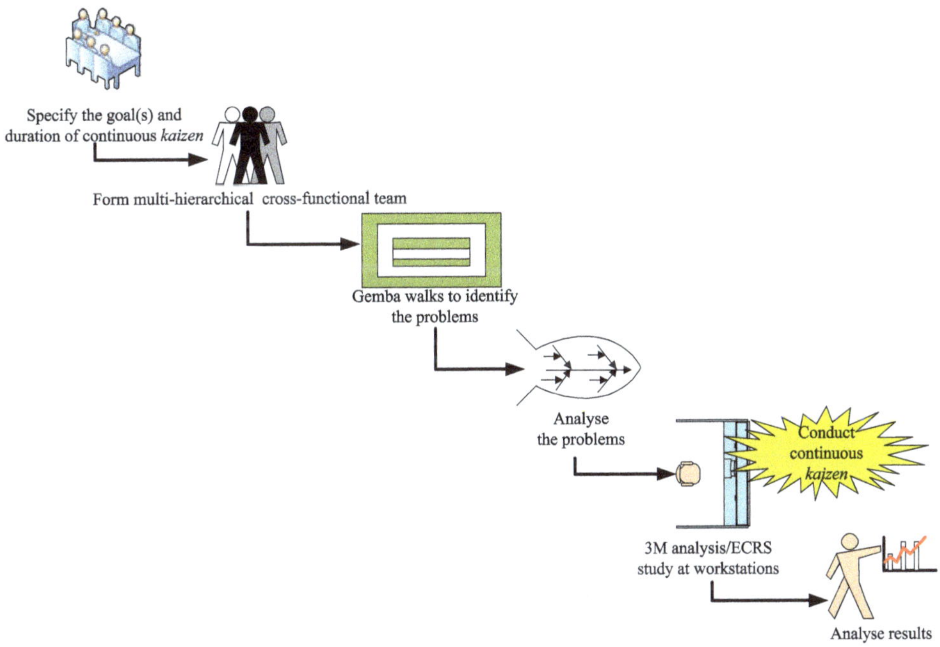

Fig. 6.1 Research methodology for continuous kaizen

6.4.1 *Case Organization*

The ABCL (organization identity is hidden for the confidentiality) is one of the prominent multinational automotive component manufacturer of India. ABCL has received the TPM excellence award in the year 2007 and Deming prize in the year 2003 and possesses ISO 9001, ISO 14001 and TS 16949 certifications. The ABCL has seven plants all over India and the case study is conducted in one of these seven plants established plant in the year 2011. The plant has various machining and assembly lines—idler machining, rack housing, idler assembly, steering column assembly, rack and pinion sub-assembly, universal joint (UJ) assembly, and intermediate shaft assembly line. The top management agreed to implement the 'continuous kaizen' as a line improvement project. Since, the 'continuous kaizen' is a short-term project without much investment.

6.4.2 *Specify the Goals and Project Duration*

The main features of continuous kaizen project are to specify the goals or targets of improvement initiatives within a defined period. The management gave a target of 5% increase in productivity and line efficiency. The project duration was fixed as 3 months to carry out continuous kaizen activities throughout the value chain (line).

6.4.3 Form Multi-hierarchical Cross-Functional Team

A multi-hierarchical cross-functional team comprising 10 internal members and two members as external experts. The team involved four head of departments at senior manager level from production, quality, manufacturing engineering (ME), and production planning and control (PPC) departments; two team leaders (TLs) at the manager level from ME and PPC departments; two assistant team leaders (ATLs) at the supervisor or assistant manager level from quality and production departments; two operators at the operational level; and the authors of the study as external experts.

6.4.4 Gemba Walks to Identify the Problems

To analyse the present situation of the line, the team obeyed the lean concept of "walk the flow, create the flow". A series of Gemba walks were conducted to understand the various problems associated with steering column assembly line. Figure 6.2 clearly depicts that a number of processes are arranged in parallel, which create problems to understand and analyse the process flow. The steering column assembly requires 18 processes and ten operators. Since the number of processes are more than the number of operators, therefore multi-machine activities (MMA) are considered as a work cell as shown in Fig. 6.2. Figure 6.3 clearly shows that there is large variation

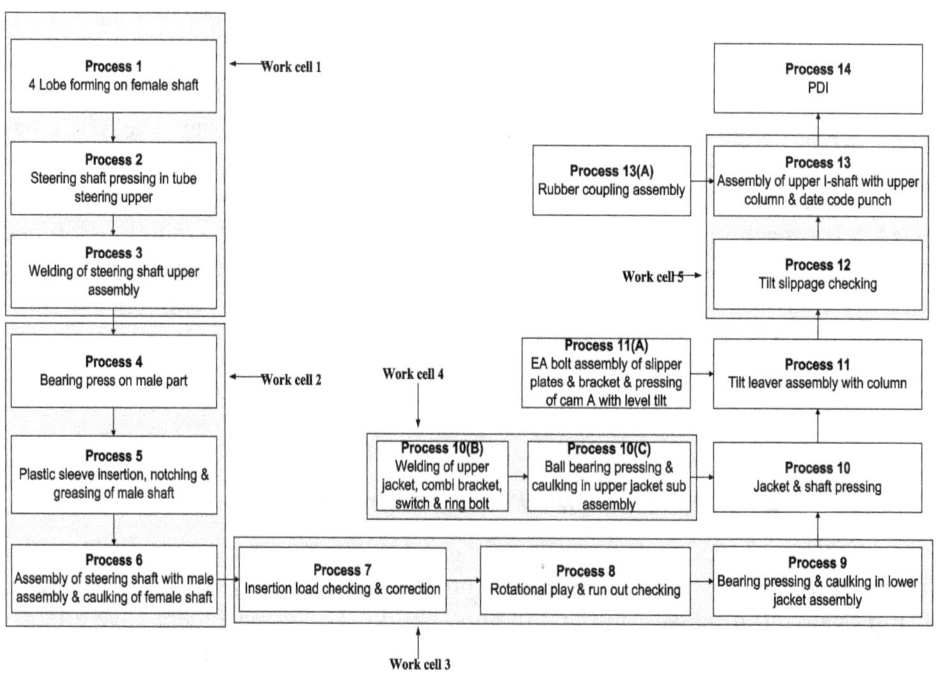

Fig. 6.2 Process sequence of steering column assembly

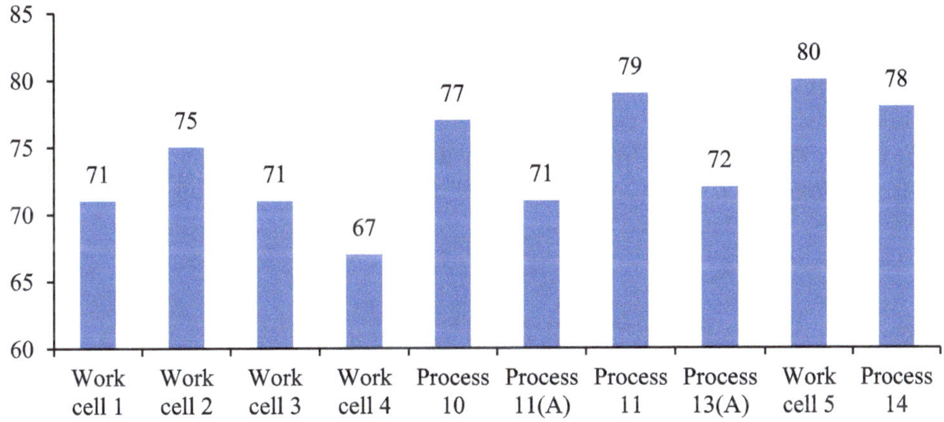

Fig. 6.3 Cycle time of the selected assembly line

in the cycle time (varies from 67 to 80 s) which unbalances the line. Similarly, the overall cycle time of the line is 80 s. The work cell 5 is the bottleneck and has the highest cycle time of 80 s, which results in low productivity.

6.4.5 *Analyze Various Lean Wastes*

Defect Waste. A fishbone diagram is developed to analyse the problem of defect waste and identify the root causes for the same. There are number of possible causes for the defect waste as shown in Fig. 6.4. The root cause for the large number of defects is poor load cell performance at process 7.

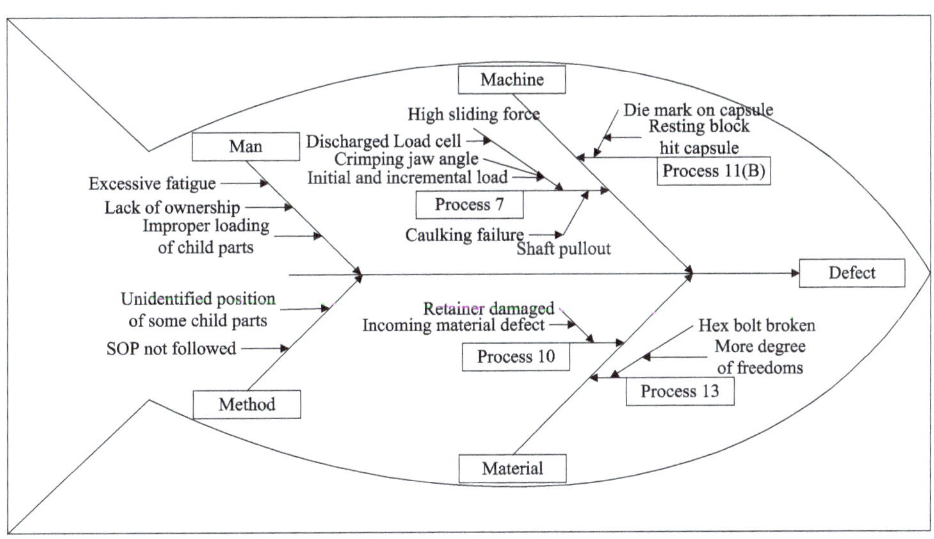

Fig. 6.4 Fishbone diagram for defect waste

WIP Inventory. The possible causes of WIP inventory waste are sorted by man, machine, material, and method causes (Fig. 6.5). The push system used to place the child parts at the various workstations is the root cause of excessive WIP inventory of child parts.

Waiting Waste. The fishbone diagram depicts the possible causes of waiting waste (Fig. 6.6). Two major causes of the waiting waste are unavailability of bought out parts (BOP) and machine breakdown.

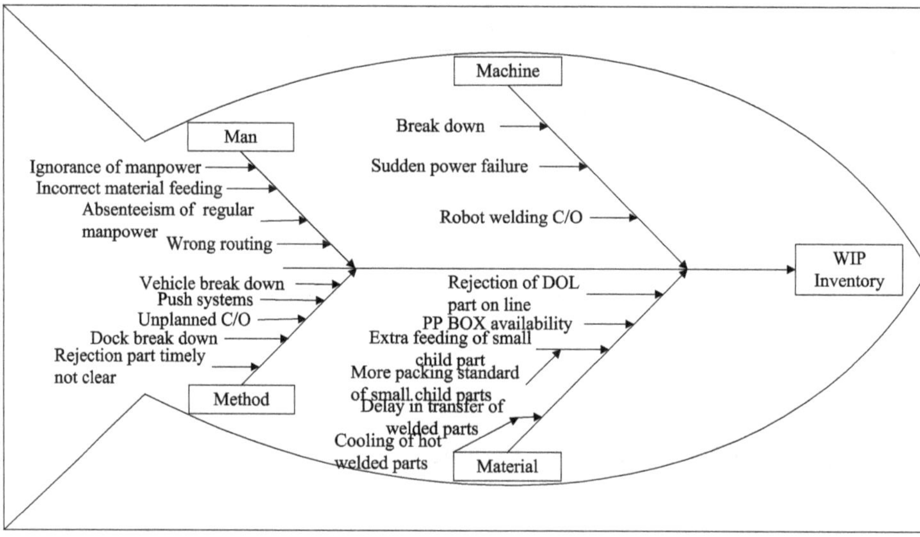

Fig. 6.5 Fishbone diagram for WIP inventory waste

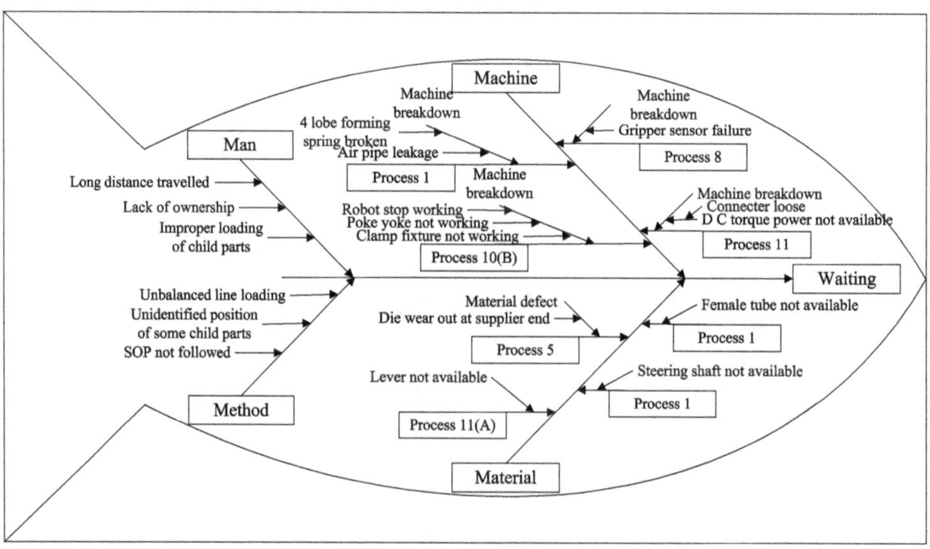

Fig. 6.6 Fishbone diagram for waiting waste

Motion Waste. The motion waste leads to high cycle time at some processes. The possible causes of motion waste are sorted by man, machine, material, and method causes (Fig. 6.7). The team found that the poor line layout and poor workstation design are the root causes of the motion waste.

Transportation Waste. The multi-hierarchical cross-functional team conducted the brainstorming session to identify and analyse the possible causes of transportation waste. Total nine possible causes are listed as presented in Table 6.1. The team concluded that the poor line layout is the root cause of transportation waste. There are a number of zigzag movements of material and operators due to the poor layout design causing transportation waste.

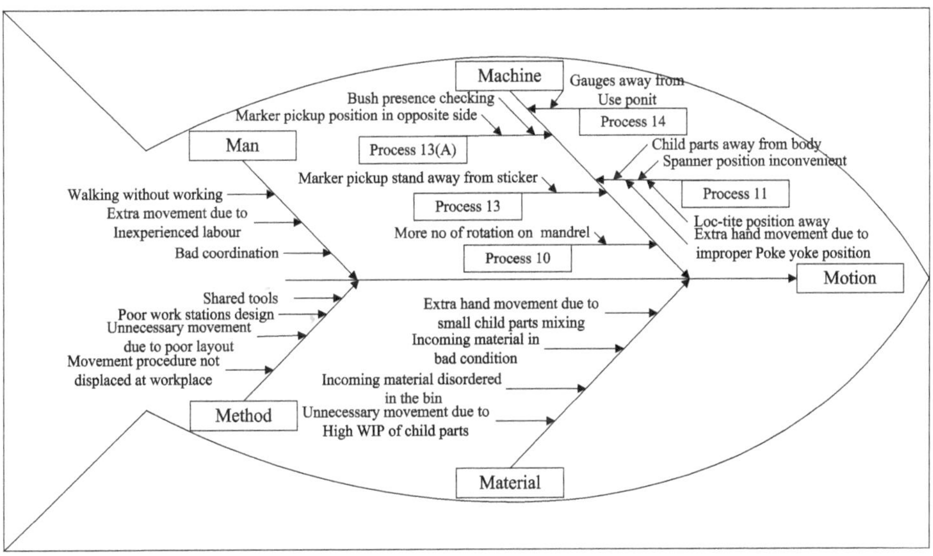

Fig. 6.7 Fishbone diagram for motion waste

Table 6.1 Possible causes of transportation waste after brainstorming session

S. No.	Possible cause
1.	Wrong routing for child part feeding
2.	More space required due to high WIP
3.	Improper bin sizes
4.	Poor line layout
5.	Poor trolley design
6.	Poor material handling
7.	Single storage location
8.	Complex working environment due to large no of child parts
9.	Poor consideration to material flow

After analysing different types of lean wastes, the team summarised these wastes in terms of occurrence of these wastes at individual processes, their root causes, possible solutions, and the improvement plans. The team categorized the improvement plans into long-term (up to one year) and short-term (0–3 months). The motion waste due to poor workstation design can be reduced by implementing the continuous kaizen at various workstations in the given project duration of three months. Thus, the team decided to implement the continuous kaizen to reduce or eliminate the motion waste at various workstations to improve line leverage by enhanced the productivity, line balancing and line efficiency of steering column assembly line.

6.4.6 Implementation of Continuous Kaizen

A series of kaizen activities are carried out at the various processes to accomplish the continuous kaizen. The team first conducted the 3M analysis and ECRS study (process 13A) to minutely understand the different activities at the individual processes.

Kaizen at Process 14. The process 14 is pre dispatch inspection (PDI) and this is the last process of the steering column assembly line. Further, this process has high cycle time of 78 s. Thus, process 14 is the pacemaker process and decides the heartbeat of the assembly line. The fishbone diagram of motion waste (Fig. 6.7) shows that the high cycle time is due to unnecessary movement of the operator. To critically analyse the situation, the cross-functional team did the 3M analysis of process 14 as presented in Table 6.2.

The team decided to implement the kaizen for the improvements of these two activities. First, the team found that the activity of 'press enter key' just after scanning the sticker code is not required and can be eliminated by changing the software of computer. The activity was eliminated by improving the software. This elimination of unnecessary hand movement decreased the activity duration from six seconds to four seconds and reduced the operator fatigue. Second, there was an unnecessary movement of operator due to poor trolley design. The cross-functional team suggested improvement in the existing trolley design. The new trolley reduced the operator movement and decreased the activity duration from nine seconds to three seconds. Thus, the cycle time of process 14 is reduced from 78 to 70 s.

Kaizen at work cell 5. Next, the work cell 5 has two processes: process 12 and process 13. To comprehend the whole activities of work cell 5, the cross-functional team carried out 3M analysis as presented in Table 6.3. After 3M analysis, the team found that the process 11 operator can carried out the two of the process 12 activities (load the part and start cycle). This reduced the cycle time of work cell 5 by four seconds. The process 13 is used to assemble upper I-shaft and upper steering column and to punch the date code on the part. The marker used to write on the sticker was far away from the writing area in the existing working condition. This kaizen reduced the hand movement from 240 to 80 mm and saved 2 s per part. For the work cell 5, total cycle time reduces from 80 to 74 s.

Table 6.2 3M analysis of process 14

S. No.	Process 14	Time (sec)	Observation	Action plan
1	Uplift lower jacket and apply cotton	3		
2	Confirmation of torque on bolt	4		
3	Down lower jacket and check lever movement	6		
4	Pick-up marker, mark on part	6		
5	Gauge checking and marking	3		
6	Apply cover on part	3		
7	Put marker and press push button to unclamp	5		
8	Scan sticker code and press enter key on the keyboard	6	Unnecessary hand movement for pressing the enter key	Eliminate the unnecessary hand movement by improving the software
9	Pick-up part from fixture and put on next fixture	4		
10	Rotation torque checking	4		
11	Thread checking using Go gauge	3		
12	Thread checking using No-Go gauge	2		
13	Serration checking using Go gauge	3		
14	Serration checking using No-Go gauge	2		
15	Flush pin checking using gauge	4		
16	Apply cover on thread	3		

(continued)

Table 6.2 (continued)

S. No.	Process 14	Time (sec)	Observation	Action plan
17	Pick-up part, put on trolley and trolley adjustment after every 6 parts	9	Poor trolley design	New trolley design
18	Come back for next cycle	3		
19	Shift loaded trolley (after 24 parts) and bring empty trolley for next loading	5		

Table 6.3 3M analysis of work cell 5

S. No.	Work cell 5	Time (sec)	Observation	Action plan
1	Unload the part	4		
2	Load the part	2	Can be accomplished by previous operator	Assign this activity to process 11 operator
3	Start cycle	2	Can be accomplished by previous operator	Assign this activity to process 11 operator
4	Put part on fixture	4		
5	Pick-up coupling and check gauge	6		
6	Insert coupling in shaft and clamp	4		
7	Pick-up bolt and washer and fit in part	5		
8	Apply torque	7		
9	Paste part number sticker	6		
10	Start cycle	2		
11	Write lever force on sticker and paste	10	Marker pick-up stand is away from the writing point	Place the marker near to the writing area
12	Pick-up polythene and rubber and apply on part	8		
13	Pick-up marker and do marking	8		
14	Unload, and load part on next bin	6		

(continued)

Table 6.3 (continued)

S. No.	Work cell 5	Time (sec)	Observation	Action plan
15	Press push button	2		
16	Pick-up next part for next cycle	4		

Kaizen at process 13(A). The process 13(A) is rubber coupling assembly. In this process, the team decided to conduct the ECRS (eliminate, combine, reduce, or shift) study instead of 3M analysis. The team observed that the process 13(A) had number of activities, which can be eliminated or combined or reduced or shifted as presented in Table 6.4. Two activities (17 and 18 in Table 6.4) were identified as poor activities. After identification of poor activities, the team decided to implement the kaizen. Three kaizen activities are carried out for improvement of the process 13(A). First, the marker pickup position was on the opposite side of the picking hand and away from the sub-assembly (part). This sub-activity can be reduced by relocating the marker on right hand side and near to the sub-assembly. This relocation of marker position reduces the operator fatigue and cycle time by 0.6 s per part. Second, the operator code is required due to WIP inventory of sub-assemblies. This sub-activity is eliminated by implementing the single piece flow. This kaizen further reduced the activity duration by 0.4 s per part. Third, the activity of 'check bush presence' required 3 s to complete. This activity is shifted to previous workstation of universal joint (UJ) assembly line. Thus, the activity of 'check bush presence' can be carried out in UJ assembly line. The team shifted this activity to previous workstation resulting in the reduction of cycle time of process 13(A) by three seconds per part.

Kaizen at process 11. The process 11 is 'tilt lever assembly with column'. The cross-functional team conducts the 3M analysis for the process 11. The 3M analysis indicated that there were numerous activities, which had the motion waste as presented in Table 6.5.

After 3M analysis, *kaizens* were conducted. First, total six types of small child parts are used to assemble the tilt lever assembly with column. The child part bins were away from the operator due to which operator had to move a long distance to pick these child parts. The child part bins were relocated near to the operator. This kaizen reduced the operator fatigue and saved five seconds per part. Second, the team conducted a kaizen and provided the inclined stand near the assembly area for easy pickup of spanner and less hand movement. This kaizen saved the activity time of 0.5 s per part. Third, cluttering of brackets and cam-lever subassemblies creates difficulties for the operator to pick the parts and operator requires extra time for the sorting. The team conducted a kaizen activity and provided a separate stand for the bracket and cam-lever subassemblies. This kaizen saved the activity time of 1.5 s per part. Fourth, the team conducted a kaizen activity and provided the loctite stand at the point of use. This kaizen saved the activity time of 0.5 s and reduced the hand movement from 260 to 210 mm. Fifth, the team conducted a kaizen activity and placed the marker nearer to the marking point (assembly area). This kaizen further reduced

Table 6.4 ECRS study of process 13(A)

S. No.	Elemental work	Time taken	Can be eliminated	Can be combined	Can be reduced	Can be shifted to next/prev. process
1	Pick-up two bolts and put on the fixture	2	×	×	×	×
2	Pick-up upper I-shaft, apply grease and put on the fixture	4	×	×	×	×
3	Pick-up spacer and fix on the fixture	2	×	×	×	×
4	Pick-up and apply rubber coupling	3	×	×	×	×
5	Pick-up stopper plate and fix on the fixture	3	×	×	×	×
6	Pick-up two nuts and apply ring gauge	5	×	×	×	×
7	Apply torque	5	×	×	×	×
8	Confirm torque by spanner	4	×	×	×	×
9	Unload part and put two bolts on the fixture	5	×	×	×	×
10	Pick-up washer and put on the part	3	×	×	×	×

(continued)

Table 6.4 (continued)

S. No.	Elemental work	Time taken	Can be eliminated	Can be combined	Can be reduced	Can be shifted to next/prev. process
11	Load the part on fixture	4	×	×	×	×
12	Pick-up and apply spacer	3	×	×	×	×
13	Pick-up UJ assembly and put on the part	3	×	×	×	×
14	Pick-up nut and apply torque	5	×	×	×	×
15	Pick-up nut and apply torque at other side	5	×	×	×	×
16	Confirm torque by spanner	4	×	×	×	×
17	Unload part, mark on part and write operator code	5	√	×	√	×
18	Check bush presence	3	×	×	×	√
19	Unload part and put on next stand	4	×	×	×	×

Table 6.5 3M analysis of process 11

S. No.	Process 11	Time (sec)	Observation	Action plan
1	Pick-up mounting bracket and put on the fixture	4		
2	Pick-up sub-assembly and fix it on fixture	7		
3	Pick-up washer and apply loc-tite	4	Pick-up time more as child parts are away from the operator	Put child part bin nearer to the operator
4	Pick-up nylon nut, apply torque by spanner and release	9	Spanner position incon-venient	Inclined stand provided for spanner
5	Pick-up cam B and fix it on part	4		
6	Pick-up lever cam assembly	4	No designated place for mounting bracket & lever	Provide stand for lever cam assembly
7	Pick-up plain washer, needle bearing and fix on part	9		
8	Apply loctite and pick-up nut	7	Loctite location is away from the assembly point	Place the loctite near to the assembly point
9	Apply torque and check lever movement	7		
10	Check rotational play	6		
11	Check lever force	7		
12	Pick-up marker and write force value	4	Marker pick-up is dis-tance long	Re-locate the marker near to marking point
13	Unlock fixture	2		
14	Unload part and put in the bin	5		

the activity time by 0.5 s and reduced the hand movement from 370 to 280 mm. Total 8 s are saved by conducting the kaizen at the process 11. However, to avoid the waiting waste, the operator working on process 11 is given two additional activities (load the part and start cycle) of process 12. Due to these additional activities, the cycle time of process 11 is increased by four seconds.

Kaizen at Process 10. Process 10 is jacket and shaft pressing. The team decided to conduct the 3M analysis to critically analyse the whole activities of process 10. The 3M analysis indicates that two activities (3 and 10 in Table 6.6) have motion waste. The team conducted a kaizen activity and reduced the number of mandrel threads by decreasing the length of the mandrel. This kaizen saved the activity time of three seconds and reduced the hand movement due to decreased number of threads from 12 to six, which also results in lesser operator fatigue.

Table 6.6 3M analysis of process 10

S. No.	Activity	Time (sec)	Observation	Action plan
1	Pick-up part from oil box with circlip assembly	3		
2	Insert in outer jacket assembly.	3		
3	Pick-up mandrel and assemble in part	6	Mandrel assembly time is high due to 12 no of threads	Reduce no of threads on mandrel
4	Load part in fixture and clamp	5		
5	Start cycle	2		
6	Pick-up retainer and assemble in outer jacket	5		
7	Pick-up part, fit circlip and mark on it	8		
8	Put part in oil box	4		
9	Unload part, de-clamp and put on fixture	6		
10	Disassemble mandrel	5	Mandrel disassembly time is high due to 12 no of threads	Reduce no of threads on mandrel
11	Pick-up circlip and assemble in part	5		
12	Pick-up marker, mark for circlip presence	3		
13	Check shaft movement	7		
14	Write load value on part and put on the next fixture	6		
15	Check slot dimension using Go-No Go Gauge	6		
16	Come back to pick-up part for cycle next cycle	3		

6.5 Results and Discussion

The case study has exhibited that considerable improvements are achieved through the implementation of continuous kaizen. The target of productivity improvement is achieved by accomplishing the continuous kaizen project. The production per labour hour (PPLH) is increased from 4.5 to 4.8 (Fig. 6.8) or the productivity is increased by 6.70%.

Correspondingly, the line efficiency is also enhanced by 2.9% as shown in Fig. 6.9. The cycle time reduction is accomplished by the elimination of motion wastes. The total cycle time is reduced from 741 to 716 s. The overall cycle time of the line is reduced from 80 to 75 s. Further, the line balancing is also improved by decreasing the cycle time variation from 4 to 2.84 σ.

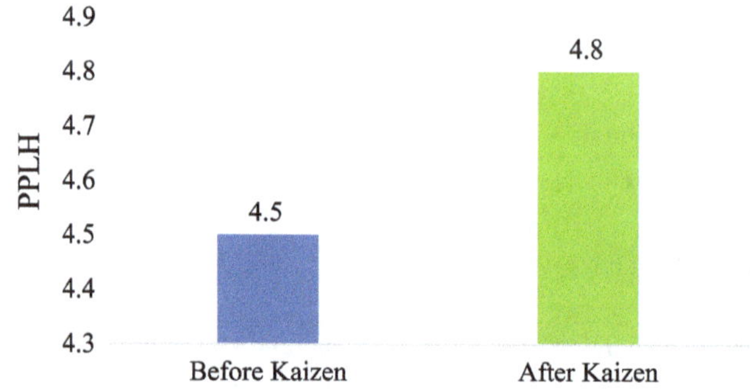

Fig. 6.8 Production per labour hour for steering column assembly

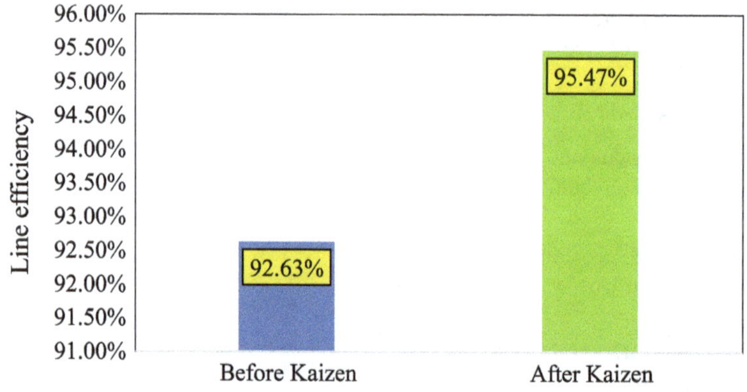

Fig. 6.9 Line efficiency of steering column assembly line

6.6 Conclusions

This paper presents a case study of Indian automotive component assembly line, which shows that continuous kaizen can be used to improve leanness of an assembly linc by improving the productivity, line balancing and line efficiency. The paper also proposes a new delineation of kaizen philosophy—continuous kaizen—which means continuous improvements at the global or whole value chain level instead of just 'change for better' at local or single workstation level. The various tools and techniques of the kaizen philosophy have been reviewed to provide salient points of each tool and technique. The paper also presents the methodology and tools for the proposed continuous kaizen project. The case study demonstrates the methodology required for the continuous kaizen. The continuous kaizen has improved the leanness of the assembly line by reducing or eliminating the motion waste to improve the productivity, line balancing and line efficiency. The total cycle time is reduced from 741 to 716 s. The overall cycle time of the line is reduced from 80 to 75 s. The continuous kaizen increased the production from 45 products per hour to 48 products per hours. The line efficiency is enhanced by 2.9%. The line balancing is also improved by decreasing the cycle time variation (standard deviation) from 4 to 2.84 σ. The continuous kaizen proves to be a versatile assembly line improvement approach facilitating the reduction of lean wastes.

References

Alukal G, Manos A (2006) Lean Kaizen: a simplified approach to process improvements. ASQ Quality Press, Milwaukee

Alvarado-Ramírez KM, Pumisacho-Álvaro VH, Miguel-Davila JÁ, Suárez Barraza MF (2018) Kaizen, a continuous improvement practice in organizations: a comparative study in companies from Mexico and Ecuador. TQM J 30(4):255–268

Aoki K (2008) Transferring Japanese kaizen activities to overseas plants in China. Int J Oper Prod Manag 28(6):518–539. https://doi.org/10.1108/01443570810875340

Bessant JR (2003) High-involvement innovation: building and sustaining competitive advantage through continuous change. Wiley

Bhuiyan N, Baghel A (2005) An overview of continuous improvement: from the past to the present. Manag Decis 43(5):761–771. https://doi.org/10.1108/00251740510597761

Chung CH (2018) The Kaizen wheel—an integrated philosophical foundation for total continuous improvement. TQM J 30(4):409–424. https://doi.org/10.1108/TQM-03-2018-0029

Farris JA, Van Aken EM, Doolen TL, Worley J (2008) Learning from less successful Kaizen events: a case study. Eng Manag J 20(3):10–20. https://doi.org/10.1080/10429247.2008.11431772

Imai M (1986) Kaizen—the key to Japan's competitive success. Random House, New York, NY

Marin-Garcia J, Garcia-Sabater J, Bonavia T (2009) The impact of Kaizen events on improving the performance of automotive components' first-tier suppliers. Int J Automot Technol Manag 9(4):362–376. https://doi.org/10.1504/IJATM.2009.028524

Marin-Garcia JA, Juarez-Tarraga A, Santandreu-Mascarell C (2018) Kaizen philosophy: the keys of the permanent suggestion systems analyzed from the workers' perspective. TQM J 30(4):296–320. https://doi.org/10.1108/TQM-12-2017-0176

Oropesa M, García J, Maldonado A, Martínez V (2016) The impact of managerial commitment and Kaizen benefits on companies. J Manuf Technol Manag 27(5):692–712. https://doi.org/10.1108/JMTM-02-2016-0021

Ramadani V, Gerguri S (2011) Innovations: principles and strategies. Strateg Chang 20(3–4):101–110. https://doi.org/10.1002/jsc.888

Salzman RA (2002) Manufacturing system design: flexible manufacturing systems and value stream mapping. Massachusetts Institute of Technology

Suárez-Barraza MF, Lingham T (2008) Kaizen within Kaizen teams: continuous and Process improvements in a Spanish municipality. Asian J Qual 9(1):1–21. https://doi.org/10.1108/15982688200800001

Suárez-Barraza MF, Ramis-Pujol J (2010) Implementation of Lean-Kaizen in the human resource service process: a case study in a Mexican public service organization. J Manuf Technol Manag 21(3):388–410. https://doi.org/10.1108/17410381011024359

Suárez-Barraza MF, Ramis-Pujol J, Kerbache L (2011) Thoughts on Kaizen and its evolution. Three different perspectives and guiding principles. Int J Lean Six Sigma 1(2):107–129. https://doi.org/10.1108/20401461111189407

Vonk J (2005) Process improvement in business permits through Kaizen. Spectrum 78(2):33–34

Yang Y, Lee PKC, Cheng TCE (2016) Continuous improvement competence, employee creativity, and new service development performance: a frontline employee perspective. Int J Prod Econ 171:275–288. https://doi.org/10.1016/j.ijpe.2015.08.006

7

Wind Speed and Temperature Forecasting using ARIMA Model

Sumanta Pasari and Aditya Shah

Abstract Due to the rapid pace of industrialization and growing demand for energy consumption, forecasting of renewable energy has become an inevitable focus of many recent studies. In this paper, our aim is to develop a univariate auto-regressive integrated moving average (ARIMA) model to forecast daily and monthly wind speed and temperature based on 15 years (2000–2014) of hourly data at Charanka Solar Park, Gujarat. To check the stationarity of time series, Dickey fuller test and rolling statistics plots are employed. Autocorrelation and partial autocorrelation plots are used to determine potential models, whereas Akaike information criterion (AIC) and Bayesian information criterion (BIC) are utilized to establish ARIMA (2, 1, 2) model. After rigorous training, model performance is validated using root means square (RMS) errors. The entire methodology is implemented in python using *pandas* for data exploration, and *stats* and *scikit-learn* libraries for model building and validation. On comparing results based on the log-likelihood, AIC and BIC values, we conclude that the ARIMA model provides better accuracy to the wind power forecasting as compared to solar power on the selected dataset.

Keywords Renewable energy · Forecasting · ARIMA

7.1 Introduction

The renewable energy sources like solar, wind, geothermal, ocean and biomass energy provide clean and replenishable energy, principally different from fossil fuels in terms of their diversity, abundance and potential to withstand energy shortage issues faced by the developing economies (Rather 2018; Kumar et al. 2010). Above all, these renewable sources support a more sustainable future by producing neither greenhouse

---

S. Pasari (✉) · A. Shah
Department of Mathematics, Birla Institute of Technology and Science, Pilani Campus, Pilani 333031, RJ, India
e-mail: sumanta.pasari@pilani.bits-pilani.ac.in; 86.sumanta@gmail.com

A. Shah
e-mail: f2014715p@alumni.bits-pilani.ac.in

gases which cause climate change nor carbon emissions (Kumar et al. 2010). There are different forms of renewable energy, depending on sunlight in a direct or indirect way. Renewable energy, by 2040, is projected to supply equal amount of electricity generation as obtained from coal and natural gas (Rather 2018; Kumar et al. 2010; Bhatia and Gupta 2018).

Solar energy is the direct conversion of sunlight (solar radiation) using solar panels or photovoltaic thermal solar collectors. It is the most abundant permanent energy resource on Earth. India, due to its geographical location, counts about 250–300 clear and sunny days in a year receiving about 4–7 kWh of solar radiation per square meter per day (Kumar et al. 2010). In particular, the western part of India experiences the highest amount of annual radiation energy. Therefore, the development of photovoltaic and solar panels is rapidly growing in India accounting for an installed capacity of 28.18 GW (as of March 2019) against an ambitious target of 100 GW by the end of 2022 (Rather 2018; Kumar et al. 2010; Bhatia and Gupta 2018). With substantial investment in solar electricity generation capacity, the cost of solar power is expected to fall about Rs. 1.90–2.30 per kWh by 2030 (Bhatia and Gupta 2018).

Wind energy, like solar power, is another abundant energy resource. However, the mechanism of wind generation is complex involving Earth's rotation, intensity of solar heat from sun, cooling effects of oceans, differential heating of the Earth's surface and the presence of other physical obstacles such as forest or mountains (Kumar et al. 2010). As a result, the wind speed at a location significantly varies over different seasons in a year. The wind power generation in India accounts for a total installed capacity of 36.63 GW (as of March 2019) against a target of 60 GW by 2022. Within a decade, the wind power cost will reduce to about Rs. 2.30–2.60 per kWh, while the storage cost will also come down by about 70% (Rather 2018; Kumar et al. 2010; Bhatia and Gupta 2018).

Apart from the solar and wind energy, biomass energy is a prominent source to meet the energy demand in developing countries. It contributes to about 32% of total primary energy, serving more than 70% of the country's population (Rather 2018; Kumar et al. 2010; Bhatia and Gupta 2018). Currently, there is about 5 GW capacity biomass powered plants against a national target of 10 GW installed biomass power by 2022 (Bhatia and Gupta 2018). Other sources of renewable energy include hydro-electricity and geothermal energy. While hydropower energy converts kinematic or potential energy of water into mechanical energy, geothermal energy is generated from the heat stored in the Earth, or the underground absorbed heat accumulation (Rather 2018; Kumar et al. 2010; Bhatia and Gupta 2018).

There have been several initiatives to forecast renewable energy variable resources, such as solar, wind and tidal using regressive models, artificial intelligence (AI) techniques, remote sensing models and numerical weather predictions (Inman et al. 2013; Lei et al. 2009; Kavasseri and Seetharaman 2009; Cadenas and Rivera 2010; Liu et al. 2012; Shukur and Lee 2015; Zhang et al. 2015; Wang et al. 2011; Tran 2013). Regressive models for day to year ahead forecasting include autoregressive (AR), moving average (MA), ARMA, ARIMA and fractional ARIMA (f-ARIMA) (Inman et al. 2013; Lei et al. 2009; Kavasseri and Seetharaman 2009).

Hybrid methods, such as ARIMA-ANN and ARIMA-Kalman are also used for wind speed prediction (Cadenas and Rivera 2010; Liu et al. 2012; Shukur and Lee 2015). In the present study, our objective is to develop ARIMA models for wind speed and solar energy (temperature) forecasting. Other concerns like cost-benefit analysis of new plants or technology development (Zhang et al. 2015; Wang et al. 2011) are not within the scope of this work.

7.2 Data Description

The data of Charanka Solar Park (23.95° N, 71.15° E) in Gujarat is procured from National Solar Radiation Database, maintained by National Renewable Energy Laboratory (NREL) (NREL homepage 2019). The obtained data comprises hourly data from year 2000 to year 2014 of the following variables: DHI (Diffuse Horizontal Irradiance), DNI (Direct Normal Irradiance), GHI (Global Horizontal Irradiance), clear-sky DHI, clear-sky DNI, clear-sky GHI, dew point, temperature, pressure, relative humidity, solar zenith angle, precipitable water, snow depth, wind direction and wind speed. For this study, we analyze the variables temperature and wind speed independently as univariate time series.

7.3 Methodology

Unlike traditional energy forecasting using probability distributions, here we develop a linear Autoregressive Integrated Moving Average (ARIMA) model that uses computer programming to provide reliable results with low computational complexity (Liu et al. 2012; Shukur and Lee 2015). The ARIMA model is a generalization of ARMA model comprising AR and MA parts. The AR part indicates that the changing variable regressed on its own lagged (i.e., prior) values, whereas the MA part incorporates the dependency between an observation and a residual error from a moving average model applied to prior observations. Here the wind speed and temperature hour-wise data are represented as individual univariate time series. Using the data of 2000–2013, results are first validated for the year 2014. Similarly, using the data till 2014, we can forecast for 2015 and so on. The method of training and validation for one-year ahead in itself is flexible for the available data.

As ARIMA models can be implemented only on stationary time series, the first task is to determine whether the time series corresponding to wind speed or temperature data is stationary or not. This in turn requires to verify whether the statistical properties such as mean, variance and autocorrelation are all constant over time (Tran 2013). A non-stationary time series may be converted to a stationary time series through differencing or detrending. The Dickey-Fuller (DF) test of stationarity and rolling (moving) statistics plots are often employed for this purpose (Tran 2013; Adhikari and Agrawal 2013). While the DF test is a statistical technique to check stationarity, rolling statistics is more of a visual way to plot the mean and standard deviations over time. The DF test comprises a test statistic and some critical values corresponding to different confidence levels. To check the stationarity, the observed value of the test statistic is compared with the critical value to reject the null hypothesis (Tran 2013; Adhikari and Agrawal 2013). In brief, the regression model for the DF test uses the following equation:

$$y_t - y_{t-1} = (\rho - 1)y_{t-1} + u_t \tag{7.1}$$

where y_t is the variable of interest, t is the time index, ρ is a coefficient and u_t is the error term. A unit root is present if $\rho = 1$ and in that case, the time series becomes non-stationary. Thus, the null hypothesis for the test is H_0: $\rho = 1$ against the research hypothesis H_1: $\rho \neq 1$. In case of non-stationarity, the data may contain two factors: trend and seasonality. While trend is the variation in mean over time, seasonality is the variation between time frames (seasons). During the course of modeling, we estimate both quantities and eliminate them, if necessary (Tran 2013).

The basic technique to eliminate trend is to *log-transform* the series. This will result in penalizing larger values more than the smaller values affecting the trend accordingly. Then, to remove the trend, the noise is removed through smoothing, aggregation or polynomial fitting. This results in averaging out the trend which could be subtracted from the log-transformed time series. *Differencing* is another popular technique to eliminate (reduce) trend and seasonal effects by removing changes in the level of a time series. The third technique often used to reduce trend and seasonal components is to *decompose* the time series into various terms comprising the ones contributing to trend, seasonality, cyclic nature, irregularity and the residuals. Thus, decomposition method provides access to the residuals which are nothing but the time series after removing other components (Tran 2013; Adhikari and Agrawal 2013).

Renewable energy prediction using non-seasonal ARIMA (*p, d, q*) model is based on the non-negative integer parameters *p*, *d* and *q* where *p* is the order (number of time lags) of the AR model, *d* is the degree of differencing (for non-seasonal differences) and *q* is the order of the MA model (Tran 2013). For computing *p* and *q* values, two graphs, namely autocorrelation function (ACF) and partial autocorrelation function (PACF) are often employed (Tran 2013). The ACF is the metric to determine the correlation of a signal with its own copy delayed or separated by several time lags. Although ACF turns out to be an excellent tool in identifying the order of an MA (*q*) process, it is not very useful to identify the order in an AR (*p*) process. The PACF provides partial correlation of a stationary time series with its own prior values, and regressed the values of the time series at all shorter lags (Tran 2013; Adhikari and Agrawal 2013). The PACF is useful to identify the order *p*.

The above methodology is implemented in python using *pandas* for data exploration, and *stats* and *scikit-learn* libraries for model building and validation. The next section summarizes various results of the study.

7.4 Results and Conclusions

The rolling mean plots of temperature and wind speed data for a period of 365 days are provided in Fig. 7.1a, whereas the results of the DF test is summarized in Table 7.1. The decomposition of wind speed data into trend, seasonality and residuals is presented in Fig. 7.1b. Similar decomposition can be carried out for the temperature data.

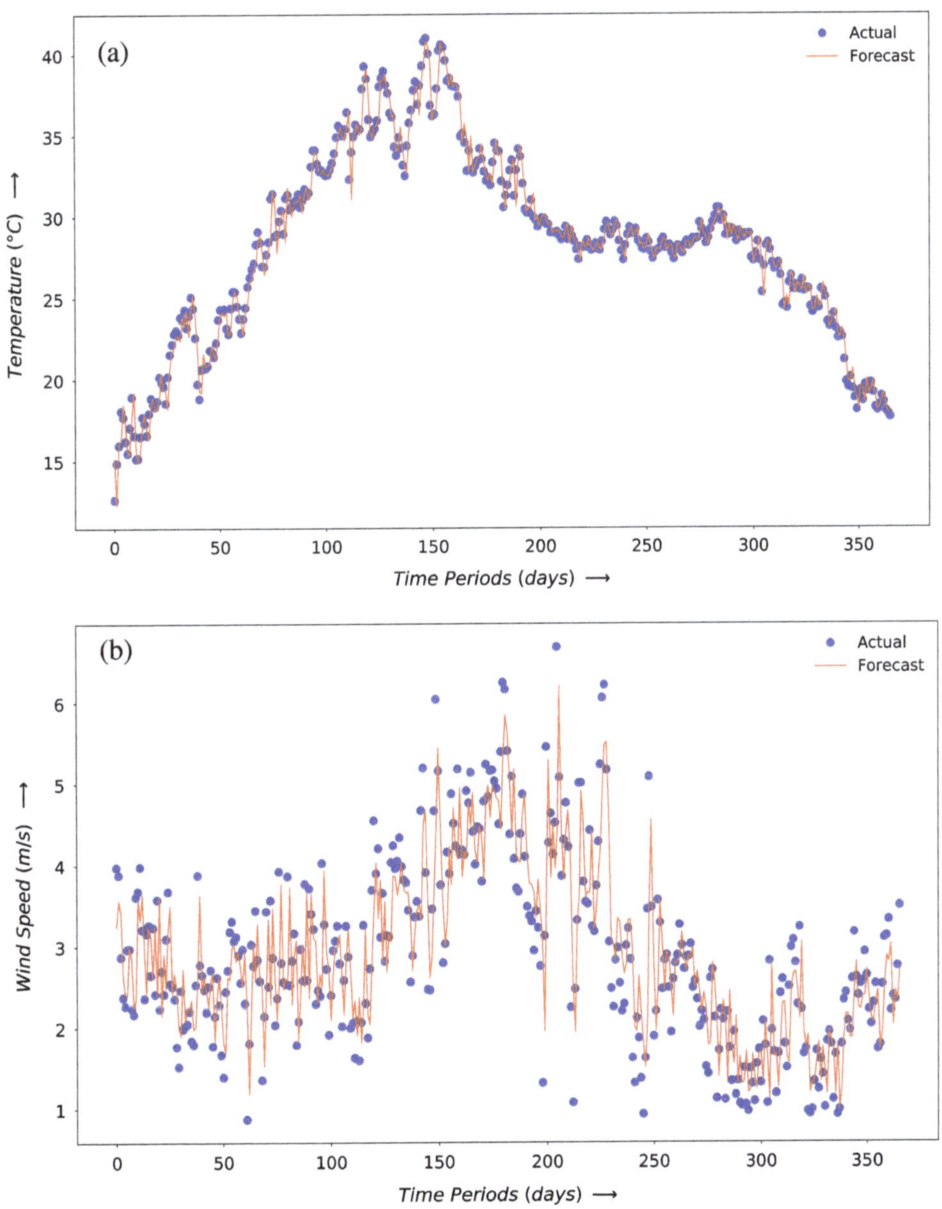

Fig. 7.1 Forecast versus actual data curve for **a** temperature and **b** wind speed data

Table 7.1 Results of Dickey-Fuller test

	Temperature	Wind speed
Test statistic	−6.96	−6.11
$p-$ value	9.39e–10	9.40e–08
Number of lags	32	26
Number of observations	5442	5448
Critical value (1%)	−3.43	−3.43
Critical value (5%)	−2.86	−2.86
Critical value (10%)	−2.57	−2.57

It is observed from Table 7.1 that for both the series, the test statistic value falls in the rejection region of the null hypothesis. Therefore, we may consider both time series to be stationary. From the rolling mean/variation plots (available on request), it is also visually observed that both time series data are stationary. In addition, the ACF and PACF plots (available on request) suggest p and q values to be 2 for both temperature and wind speed data. Thus we use ARIMA (2, 1, 2) model to forecast the renewable energy data. From the results based on the log-likelihood, AIC and BIC, we observe that the linear univariate ARIMA model provides a better fit to the observed wind speed data in comparison to temperature data.

The classification model provides a validation accuracy of 83.24%. The root mean square (RMS) errors corresponding to observed and modeled temperature and wind speed data for the year 2014 are 0.893 and 0.659, respectively, indicating a reasonable match to the observed and modeled results (Fig. 7.1).

In summary, the present study has provided a scheme to model, forecast and validate temperature and wind speed data using a univariate linear ARIMA (2, 1, 2) model. The approach is generic, scalable and suitable to be applied for other renewable energy resources for planning, unit commitment analysis and integration of renewable energy to the main power grids. Here we have forecasted one-year ahead, though the method can be easily modified to two-year forecasting with more number on input data. The method is computationally inexpensive and provides good accuracy. In future, multivariate ARIMA models or Recurrent Neural Networks (RNN) could be further explored to improve the model accuracy.

References

Adhikari R, Agrawal RK (2013) An introductory study on time series modeling and forecasting. Lambert Academic Publishing, pp 1–76

Bhatia SC, Gupta RK (2018) Textbook of renewable energy. Woodhead Publishing, India, pp 1–334

Cadenas E, Rivera W (2010) Wind speed forecasting in three different regions of Mexico, using a hybrid ARIMA-ANN model. Renew Energy 35(12):2732–2738

Inman RH, Pedro HTC, Coimbra CFM (2013) Solar forecasting methods for renewable energy integration. Progress Energy Combus Sci 39(6):535–576

Kavasseri RG, Seetharaman K (2009) Day-ahead wind speed forecasting using f-ARIMA MODELS. Renew Energy 34(5):1388–1393

Kumar A, Kumar K, Kaushik N, Sharma S, Mishra S (2010) Renewable energy in India: current status and future potentials. Renew Sustain Energy Rev 14:2434–2442

Lei M, Shiyan L, Chuanwen J, Hongling L, Yan Z (2009) A review on the forecasting of wind speed and generated power. Renew Sustain Energy Rev 13(4):915–920

Liu H, Tian HQ, Li YF (2012) Comparison of two new ARIMA-ANN and ARIMA-Kalman hybrid methods for wind speed prediction. Appl Energy 98:415–424

NREL homepage. https://www.nrel.gov/, Accessed 24 June 2019

Rather NUR (2018) Introduction to renewable energy technologies in India. Education Publishing, pp 1–140

Shukur OB, Lee MH (2015) Daily wind speed forecasting through hybrid KF-ANN model based on ARIMA. Renew Energy 76:637–647

Tran VL (2013) stochastic models of solar radiation processes. Ph.D. Dissertation. University of Orleans, France

Wang X, Guo P, Huang X (2011) A review of wind power forecasting models. Energy Procedia 12:770–778

Zhang Y, Beaudin M, Taheri R, Zareipour H, Wood D (2015) Day ahead power output forecasting for small-scale solar photovoltaic electricity generators. IEEE Trans Smart Grid 6(5):2253–2262

8

Recycling of Plastic Waste Material: 3-CYCLE Project

Max Juraschek, Lennart Büth, Sebastian Thiede, and Christoph Herrmann

Abstract Closed-loop manufacturing enables the integration of waste material and end-of-life products into the manufacturing processes of new products. As additive manufacturing in the form of filament-based 3D printing is becoming more frequently used, this process can be utilized to create a closed-loop process chain for the recycling of plastic waste material. Within the 3-CYLCE project, a modular process chain was developed for treating end-of-life products and extruding printable filament out of shredded waste material. This filament can then be utilized to create new products with 3D printers substituting virgin material. The implementation of the process chain in learning environments is discussed and the possible learning content presented.

Keywords Learning factories · 3D printing · Circular economy

8.1 Motivation

Environmental pollution caused by plastic waste has become a global challenge originating in the industrial mass production of plastic products. Global supply chains and large-scale manufacturing of products are responsible for a considerable impact on the environment (Duflou et al. 2012). Particularly in developing and emerging countries, where the disposal infrastructure has not been able to keep pace with the development of consumption and waste import, the environment has been severely polluted by large quantities of plastic waste evident also in the rise of ocean plastics (Ostle et al. 2019). Rarely, the waste material flows are recycled or reused. Plastics that are relieved to the environment are not used for energy recovery of the embodied energy in the plastic.

M. Juraschek (✉) · L. Büth · S. Thiede · C. Herrmann
Chair of Sustainable Manufacturing and Life Cycle Engineering, Institute of Machine Tools and Production Technology (IWF), Technische Universität Braunschweig, Langer Kamp 19b, 38106 Braunschweig, Germany
e-mail: m.juraschek@tu-braunschweig.de

At the same time, desktop 3D printers are increasing in popularity as they become cheaper and refined (Lee et al. 2007). The global market size for additive manufacturing is reported at 13bn USD in 2018 and projected to rise to more than 20bn USD in 2020 (Wohlers associates projected: global additive manufacturing market size between 2016 and 2020 n.d). Additive manufacturing can be an enabler for mass personalization and the introduction of distributed production systems (Rauch et al. 2016). FDM 3D printers offer an accessible, safe and easy to use generative process at comparable low investment and operating costs. A high share of learning factories offers 3D printers, leading to the opportunity to experience the life cycle of products and closed loop manufacturing (Despeisse et al. 2017). A learning factory is a close representation of a real production environment for education, research and training purposes. (Abele et al. 2015).

In engineering education and especially in learning factories, 3D printers are increasingly used for demonstration and teaching on digital fabrication technology as well as for new design workflows. At the same time, FDM 3D printing can enable a new way of recycling plastic waste by using it to manufacture new products (Muschard and Seliger 2015). Using a wire (filament) consisting of one type of thermoplastic, these printers create three-dimensional components from a digital template created on the computer. In the learning factory at the Institute for Machine Tools and Production Engineering (IWF) at Technische Universität Braunschweig, Germany, in collaboration with BITS Pilani, India, a recycling process based on an FDM 3D printer was created allowing value adding activities with waste plastics as input material.

8.2 The 3-CYCLE Process

8.2.1 Circular Material Flows and Closed Loop Production Systems

In mass-scale, centralized production systems, economy of scale effects are utilized to manufacture a high number of products in an economically efficient way. This approach inhibits several negative economic, environmental and social impacts originating in the concentration of the physical activities at one location and leading to long distances to customers. Decentralized manufacturing systems can perform better regarding these impacts under suitable circumstances (Rauch et al. 2016). One step further, realizing localized closed loop production systems can be a strategy towards reducing the environmental impact of manufacturing activities. With the utilization of waste material for the generation of new products, energy and resources consumption within manufacturing systems can be reduced if additional demands for waste material processing do not exceed these of the original raw material provision (World economic forum: driving sustainable consumption: closed loop systems 2009). The implementation of distributed production systems can furthermore strengthen local

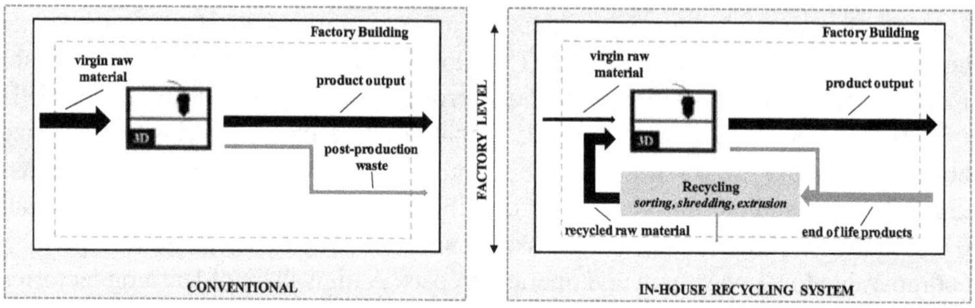

Fig. 8.1 Changed material flow with the implementation of in-house material recycling in a production system (Juraschek et al. 2016)

value creation and might have positive social impacts. A possible material source consists of end of life products from thermoplastics that lost their value (Ashby 2013). The physical, functional, technical, economical and legal characteristics determine the lifetime of a product (Ashby 2013). The nature and characteristics of the reverse flows can be classified in three groups (Juraschek et al. 2017):

- Returns during the use phase: products are usually slightly used (i.e. returned before the first month) and in most cases still state of the art technology
- Returns after the use phase: products that have been intensively used and their technology is outdated
- Returns after the End of Life: products with no competitive functionality (e.g. old equipment and products that need high amounts of energy and resources).

Three general approaches can be applied for end of life products to reinsert residual value: re-use, remanufacturing and recycling (Ashby 2013). The latter one, recycling, is defined as the reprocessing of recover material at the end of life. According to Ashby (Ashby 2013) it can be the best alternative to recover value from waste streams.

The processes for recycling are commonly a series of steps aiming at separating the different materials found in a product. Remanufacturing on the other hand is "the restoration of a used product to like-new condition with respect to quality by replacing components or reprocessing used parts" (Lund and Mundial 1984). Re-use of products requires the return of the product to a further using phase. Utilizing these approaches, negative economic and environmental impacts can be avoided in many cases as compared to disposal or incineration. With the implementation of recycling processes in a production system, the material flows can change significantly as illustrated in Fig. 8.1.

8.2.2 Utilization of FDM 3D Printers for Closing the Loop

Fused Deposition Modeling (FDM) 3D printers commonly use plastic material in the form of a filament as input for the production process. If the virgin material used

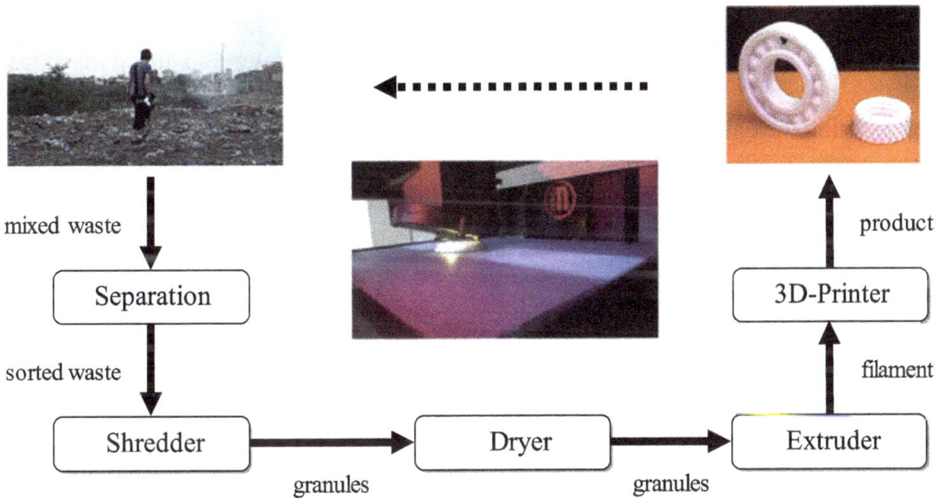

Fig. 8.2 Schematic flow of the 3-CYCLE process for plastic materials

for the generation of the filament is substituted by waste material, the environmental impacts of the products system can be lowered (Juraschek et al. 2016). 3D printing can be favorable over conventional manufacturing processes, as even complex products can be produced in small quantities. To transform the waste plastic into filament, a process chain is required consisting of a shredding and separation process for plastic waste, an extrusion process for the production of the filament and a 3D printer (see Fig. 8.2).

As part of 3-CYCLE,[1] an international student group was established to investigate the recycling process technologically, environmentally and economically as well as to exchange ideas with Indian students in order to realize a meaningful utilization of the process. The existing cooperation of Technische Universität Braunschweig and BITS Pilani in the Indian state of Rajasthan and the infrastructure of the institutes were used as a starting point and expanded in the course of the project. However, for defining the requirements of a recycling process that does not only work in a laboratory environment, the conditions at both sites had to be investigated. In order to optimize the process to the local requirements, information about the available plastic quantities and types were collected and interviews conducted, resulting in the decision of using on Polyethylene terephthalate (PET) as material, as it is widely available in waste flows in India and Germany.

The learning factories were used to perform extensive tests on the extrusion of PET. These experiments helped to identify the right extruder type. A screw extruder and a piston extruder were compared. The screw extruder is a continuous process that allows the continuous extrusion of filaments. The piston extruder is a batch process. The piston extruder has proven to be advantageous for the extrusion of PET and larger material fragments while drastically reducing the need for preparatory work (crushing of the material). However, this again has disadvantages with regard

[1]The project name refers to the goal to recycle plastic material with 3D-printing.

to an additional necessary process step of drying the material. The screw extruder experienced frequent blockages by the inconsistent waste material. During the extrusion process, challenges regarding the moisture in the material and with the resulting vapor during the extrusion process occurred. In order to achieve a uniform result with regard to the filament diameter, a cooling section was developed for the extruded filament before it is wound up. A favorable source for waste material are PET bottles. For processing, the bottles were separated from caps and labels, cleaned, crushed and pre-dried in an oven. PET is a hygroscopic material and if the water content in the material is too high, bubbles are thrown by evaporating water during the extrusion process. Direct pre-drying in the extruder at low extruder temperatures has proven to be advantageous, as the material no longer has any chance of absorbing new water in the atmosphere. The drying process in the extruder takes at 120 °C four hours. The homogenization process in the extruder for the material at 200 °C requires approximately one hour. These values were determined empirically based on several experiments. The dry and warm ambient conditions in India turned out to be an advantage and allowing reduced drying time compared with Germany.

8.3 Implementation in Learning Factories

For the implementation in the learning factories, several learning modules were created based on the 3-CYCLE process as shown in Table 8.1. The learning content is enabled by the physical process set-up and allows project-based and self-guided learning.

Additive Manufacturing and Direct Digital Production Technologies. With the 3-CYCLE process, learners are able to experience an additive manufacturing

Table 8.1 Potential learning contents of the 3-CYCLE process in learning factories

Module	Learning content (Selection)
Manufacturing technology	• Additive manufacturing and FDM 3D printing process • Design and process workflow • Influence of printing and filament parameters • Micro controllers and sensors
Circular material flows and assessment	• Life Cycle Assessment of 3D printed products • Analysis of material and energy demand • Identification of significant parameters and hot spots
Material recycling and sorting	• Material contamination and sorting • Ageing of material over recycling cycles • Material testing
Assessment of new technologies	• Structured testing and assessment of technology innovations • Integration of new devices into existing processes

process and the generation of the filament required. They can change parameters, influence the filament quality and evaluate the printing outcome. This imparts an understanding of the FDM process, its potentials and challenges. The new digital design and process workflow from CAD model to printed products is also made accessible. The filament extrusion process itself offers several learning fields as shown in Fig. 8.3. The process is set up with different learning activities prepared. Information and labels for the process flows are removable and students can elaborate and test their knowledge in a self-guided way. The integrated micro controllers and sensors can be manipulated. With the built-in quality monitoring, production waste and yield losses can be investigated.

Circular Material Flows and Assessment. The environmental performance and implications of closed loop production systems can be practically investigated with the 3-CYCLE process. Practical learning activities include data acquisition on material and energy flows and the exploration of the influence of process parameters. Further learning content is a module on life cycle assessment of 3D-printed products and the comparison of the closed loop system with a printing process based on

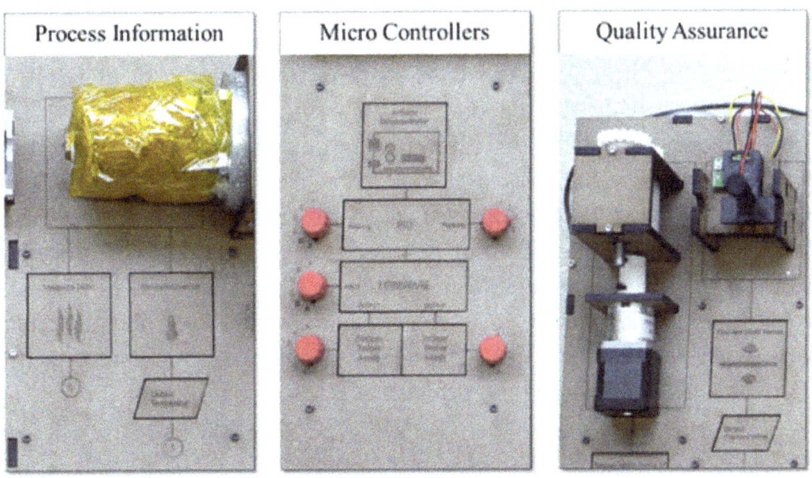

Fig. 8.3 Exemplary learning fields for the recycling process as implemented in the learning factory at TU Braunschweig for engineering education

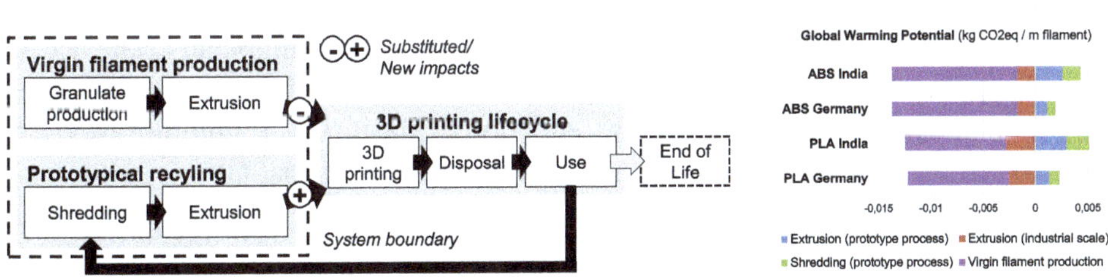

Fig. 8.4 Framework of screening LCA for the 3-CYCLE process for assessment of the environmental performance and screening LCA results for a case study (Juraschek et al. 2016)

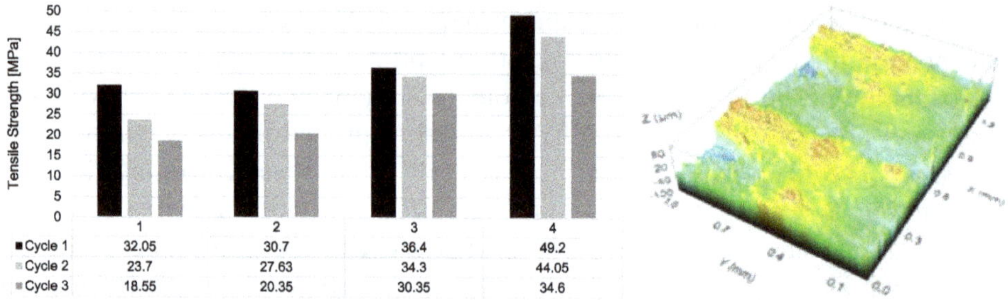

Fig. 8.5 Investigation on the ageing of the plastic material regarding strength over several recycling cycles (left) and exemplary investigation of surface quality (right)

virgin raw material. With a screening LCA approach shown in Fig. 8.4, significant parameters and environmental hot spots can be identified.

Material Contamination and Ageing. In recycling processes, purity and quality of the acquired waste material is of high importance. This can practically be demonstrated with the 3-CYCLE process. Learners are able to try different sorting techniques for plastic materials and compare their performance. Furthermore, the ageing of the plastic material over the recycling cycles can be investigated. By printing, testing and recycling iteratively the same material, degradation of material quality can be analyzed as results from a student project illustrate in Fig. 8.5. Exemplary results have shown that within in three cycles the material strength degraded to 58–83% of its original value.

Assessment of New Technologies. Due to its modular design, the 3-CYCLE process allows the implementation of new technologies. For instance, a handheld near-infrared spectroscopy device was introduced to the material sorting process and tested regarding its performance for material identification. Learners are enabled to practically experience methods for structured testing and assessment of technology innovations in a near-industrial environment of a learning factory.

8.4 Conclusions

The 3-CYCLE process chain was developed and evaluated in the environment of learning factories and constructed from robust and simple components. The access to such a process opens up many potentials. In addition to the positive effects on people and the environment through the reduction of plastic waste, complex products can also be developed and manufactured with simple means that allow value generation from waste. Furthermore, there is great potential for teaching on the subject of sustainability in manufacturing systems and closed loop production systems. Several influential factors were identified during the project for the success of the 3-CYCLE process (Juraschek et al. 2017):

- The equipment is required to be of a robust nature and there are only a limited number of machines commercially available on the required scale.
- The parameters of the extrusion process and the quality of the waste material are most critical for the resulting filament quality.
- The requirements for the waste material depend strongly on the extruder and influence all upstream processes.
- The extrusion process is the main energy demanding process and the process allowing most of the learning modules.
- The development of a continuous recycling and printing process is technically feasible although it would lead to a higher complexity in construction and operation.

A significant challenge was to find the best operating parameters for the environmental conditions at both sites in Pilani and Braunschweig. Especially differences in ambient temperature and humidity influence the process performance. The international cooperation on a scientific and student level has proven to be a successful concept.

Acknowledgements The authors express their gratitude to AKB-Stiftung funding the project "3-Cycle: Value creating plastics recycling with 3D-printers". The project would not have been possible without the effort put in by the students who were working on their student projects and theses: B. Willenbrock, C. Töppe, L. Büth, H. Khandelwal, T. Kher, T. Kaufmann, D. Luedecke & M. Schmidt, C. Gang, F. Schlösser, Y. Yao, L. Kindermann, N. Jaeschke, A. Ekbote, K. Zhou and J. Sieker.

References

Abele E, Metternich J, Tisch M, Chryssolouris G, Sihn W, ElMaraghy H, Hummel V, Ranz F (2015) Learning factories for research, education, and training. Procedia CIRP 32:1–6. https://doi.org/10.1016/j.procir.2015.02.187

Ashby MF (2013) Materials and the environment Eco-informed material choice. Butterworth-Heinemann/Elsevier, Amsterdam

Despeisse M, Baumers M, Brown P, Charnley F, Ford SJ, Garmulewicz A, Knowles S, Minshall THW, Mortara L, Reed-Tsochas FP, Rowley J (2017) Unlocking value for a circular economy through 3D printing: a research agenda. Technol Forecast Soc Change 115:75–84. https://doi.org/10.1016/j.techfore.2016.09.021

Duflou JR, Sutherland JW, Dornfeld D, Herrmann C, Jeswiet J, Kara S, Hauschild M, Kellens K (2012) Towards energy and resource efficient manufacturing: a processes and systems approach. CIRP Ann - Manuf Technol 61:587–609. https://doi.org/10.1016/j.cirp.2012.05.002

Juraschek M, Bhakar V, Posselt G, Kaluza A, Böhme S, Herrmann C, Sangwan KS (2016) Recycling of plastic waste using fused deposition modeling as a production process. In: Proceedings indo-German conference sustainability, pp 28–33

Juraschek M, Cerdas F, Posselt G, Herrmann C (2017) Experiencing closed loop manufacturing in a learning environment. Procedia Manuf 9:57–64. https://doi.org/10.1016/j.promfg.2017.04.046

Lee CS, Kim SG, Kim HJ, Ahn SH (2007) Measurement of anisotropic compressive strength of rapid prototyping parts. J Mater Process Technol 187–188:627–630. https://doi.org/10.1016/j.jmatprotec.2006.11.095

Lund RT, Mundial B (1984) Remanufacturing: the experience of the United States and implications for developing countries, vol 31. World Bank, Washington, DC

Muschard B, Seliger G (2015) Realization of a learning environment to promote sustainable value creation in areas with insufficient infrastructure. Procedia CIRP 32:70–75. https://doi.org/10.1016/j.procir.2015.04.095

Ostle C, Thompson RC, Broughton D, Gregory L, Wootton M, Johns DG (2019) The rise in ocean plastics evidenced from a 60-year time series. Nat Commun 10:1622. https://doi.org/10.1038/s41467-019-09506-1

Rauch E, Dallasega P, Matt DT (2016) Sustainable production in emerging markets through Distributed Manufacturing Systems (DMS). J Clean Prod 135:127–138. https://doi.org/10.1016/j.jclepro.2016.06.106

Wohlers associates projected: global additive manufacturing market size between 2016 and 2020 (in billion U.S. dollars) n.d

World economic forum: driving sustainable consumption: closed loop systems (2009)

9

Indian Pharmaceutical Industry: Sustainability Awareness

Monica Sharma, Umesh Chaturvedi, Govind Saran Dangayach, and Prabir Sarkar

Abstract Use of technology in this era requires organizational goals to be met more than just maximizing profit and capturing more and more market share. The aim of this research is to explore sustainability awareness in Indian pharmaceutical industry by considering social well-being along with economic and environmental aspects and attainment of key drivers. Despite moving towards 21st century, it is a fact that sustainability and related practices in Indian pharmaceutical industry is still in its infancy. Thus, looking at sustainability comprehensively, its awareness and related practices needs to be explored by incorporating them at strategic, tactical and operational level functions of a pharmaceutical organization. Once these aspects are explored, this will not only help to improve the environmental performance of an organization but also enhance managerial capability and decision-making capacity. This work has been carried out as a part of an ongoing research to identify and establish theoretical relationship between sustainability awareness, triple bottom line and key drivers. Therefore, it is expected that its practical implementation can be achieved in future with greater clarity.

Keywords Sustainability · Sustainability awareness · Pharmaceutical industry

M. Sharma · G. S. Dangayach
Department of Management Studies, Malaviya National Institute of Technology Jaipur, Jaipur 302017, RJ, India

U. Chaturvedi (✉)
Institute of Business Management, GLA University, Mathura 281406, UP, India
e-mail: umesh.chaturvedi@gla.ac.in

P. Sarkar
Department of Mechanical Engineering, Indian Institute of Technology Ropar, Rupnagar 140001, PB, India

9.1 Introduction

Achieving sustainability is one of the major concerns in the pharmaceutical industry (Amran and Ooi 2014; Agar et al. 2016; Sheldon 2016). Pharmaceutical manufacturing is complex in nature, and associated with high waste generation (Sheldon 1993) and GHG emissions. Researchers have shown greater environmental impact of pharmaceutical per kilogram production compared to basic chemicals because of the complex chemical formulation involved that also leads to higher waste per kg product, and, consequently higher fossil-fuel consumption. Cumulative energy demand is 20 times greater and GWP is 25 times higher than basic chemical product's production (Wernet et al. 2010; Cespi et al. 2015). Having its concerns on climate change (http://web.unep.org/), reported that green-house gas emissions are the main culprit and dominant factor for climate change. Thus, with a goal of exploring sustainability awareness in Indian pharmaceutical industry this work aims to establish its relation with environmental aspects, economic aspects, social aspects i.e. triple bottom line (TBL) and external forces that can act as a key driver and contribute to achieving sustainability. Raman (2006) have studied that Indian pharmaceutical corporate are still hesitant while disclosing on environmental and energy issues. Further, Goyal (2014) concluded that the disclosure index on environmental practices considering clean technology, energy consumption, environmental management etc. for Indian Pharmaceutical Industry was only 22.0 (Industry-wise disclosure index is calculated by dividing total scores attained by all the companies related to particular sector with the total maximum score that can be attained, as studied by Goyal (2014), while the highest disclosure was from Oil and Gas industry at 41.42 followed by Cement industry at 40.28. This shows that environmental reporting is one of the strongest ways to achieve sustainability since it helps monitor the environmental performance and thus aids in exploring the avenues for improvement. Most if not all, the published literature in Indian context is not adequately proposing a model to highlight sustainability awareness (Goyal 2014) and its relation with TBL and key drivers. With the perspective of managerial implementation and ongoing research, this paper presents a relationship model between sustainability awareness and factors contributing to environment, economic and social aspects, along with key drivers in context of Indian pharmaceutical industry.

9.2 Research Methods

For this study mainly methodology reported in Malhotra (2004) was followed. The same was found to be more pertinent for this particular research, among others surveyed. The adoption of such methodologies also supports the survey-based research approach because data collected through survey-based methods allows researchers to observe a real-world phenomenon in greater detail. Further, it helps to test a newly formed construct (Sihvonen and Partanen 2016). Constructs (Table 9.2) pertaining to sustainability, TBL and key drivers were identified through extensive literature review as done by authors (Peukert and Sahr 2010; Mitra 2012; Watson 2012; Lozano et al. 2016; Chaturvedi et al. 2017). For this research, a survey tool was developed on a five-point Likert scale. After its construct validation, the tool was deployed for pilot testing. 53 out of 57 responses were received since the remaining four responses were not received even after many request reminders. Reliability testing was then performed to validate the constructs in the questionnaire and to see whether indicators under each constructs behave suitably and indicate what they are meant to. The observed Cronbach's alpha values ranged from 0.853 to 0.862 which were well above the threshold value of 0.70 (Nunnally 1978), assuring internal consistency of measured items. Thereafter the survey tool was sent for final survey. The sampling frame considered all the pharmaceutical companies who were registered in the directory of National Pharmaceutical Pricing Authority (NPPA). A total of 2147 Indian pharmaceutical companies were contacted via email. Selection of these companies for establishing contact had been done through convenient, random and snowball sampling technique. Out of all the total companies contacted, responses from 439 were obtained. Out of these 439 respondents, 393 (i.e. response rate of 18.30%) were found suitable based on complete information provided against each items of questionnaire, for further analysis. The response rate was found to be similar in line with other such web-based surveys carried out by researchers like Sihvonen and Partanen (2016) with 15.5%, Lozano et al. (2016) with 8.9% and Gopal and Thakkar (2015) with 16.2% response rate. The comparison of response rates is relevant here since the response rate in industrial surveys is usually very low and hence the rate of 18.3% obtained in this work is supported by the rates reported in other similar works done for other industries elsewhere.

9.3 Research Results and Data Analysis

The Survey responses (through survey questionnaire pertaining to constructs and its indicators and Item Statistics in Table 9.3) were analyzed using descriptive and inferential statistics. The mean value from item statistics table with corresponding standard deviation ranged from 3.46 to 4.53, showing that majority of the respondents agreed with the scaled variables. After getting mean importance value of all scaled items, reliability and validity analysis was done. Cronbach's alpha values varied from 0.701 to 0.774 ensuring internal consistency between items pertaining to sustainability, TBL and key drivers. Next to reliability, KMO and Bartlett's test of sphericity were performed to check sampling adequacy. KMO measures of sampling adequacy were 0.767 for items pertaining to sustainability and 0.791 for items consisting to TBL and key drivers, thus ensuring sampling adequacy. This further confirmed that the items listed in the questionnaire were valid and relevant in terms of both the questions themselves and the study as a whole.

9.3.1 *Factor Analysis*

After reliability and sampling adequacy tests, factor analysis tests (the Kaiser criterion (Eigen Value > 1), in conjunction with evaluation of scree plot) were performed. Primary component analysis followed by varimax rotation component matrix was conducted. Value of factor loading ranges from 0.452 to 0.895 > 0.4, showing strong loadings on its original factors (Hair et al. 2006). The results of factor loading show that the variables consisting sustainability were loaded on three factors and named as sustainability awareness, sustainability planning and sustainability implementation whereas variables of environmental aspects were loaded on two factors and were named as material impact and process optimization. Variables consisting of economic aspects, social aspects and key drivers were remaining loaded on its original form. Reliability of newly formed constructs was further tested and found to be above the minimum accepted value i.e. 0.70.

Following Hypothesis was framed to test the relation between sustainability awareness and the enablers/factors identified through factor analysis.

H_1 = There is significant association between sustainability awareness and material impact, process optimization, economic aspects, social aspects and key driver.

9.3.2 *Inferential Statistics*

After factor analysis, inferential statistical analysis was performed to explore the relation between sustainability awareness and TBL along with key drivers. The correlation table (Table 9.1) shows a strong correlation between material impact and sustainability awareness ($r = 0.888, p < 0.001$), followed by process optimization ($r = 0.639, p < 0.0001$).

Results depict that awareness about impact of materials in use will help organization to optimize the whole process. In other words, we can say that once an organization knows about impact of materials and its environmental consequences, it will help engineers, strategists and managers to optimize the whole process in order to enhance its environmental performance. Once the process optimization is implemented it will reduce waste generation, ensure less emission and increased yield thus ensuring economic and social benefits. Moreover, closely linked social aspects ($r = 0.308$, $p < 0.001$) with sustainability awareness depicts that while becoming socially responsible organizations, companies have to attempt many goals simultaneously such as employee training and development, health and safety issues, employee participation in decision making etc. Additionally, globally-driven initiatives, legislative decisions, R&D facilities and mandatory reporting (Key drivers) will help organizations to improve their public corporate image for a long term. The result shows a positive association between sustainability awareness and predictors.

Table 9.1 Pearson correlation (2-tailed) between sustainability awareness and predictors

	Parameter	Factor 1	Factor 2	Factor 3	Factor 4	Factor 5	Factor 6
		Material impact	Process optimization	Economic aspects	Social aspects	Key drivers	Sustainability awareness
Factor 1	Cor.	1					
	Sign.						
Factor 2	Cor.	0.449**	1				
	Sign.	0.000					
Factor 3	Cor.	0.314**	0.175**	1			
	Sign.	0.000	0.001				
Factor 4	Cor.	0.233**	0.351**	0.281**	1		
	Sign.	0.000	0.000	0.000			
Factor 5	Cor.	0.290**	0.146**	0.553**	0.052	1	
	Sign.	0.000	0.004	0.000	0.305		
Factor 6	Cor.	0.888**	0.639**	0.268**	0.308**	0.296**	1
	Sign.	0.000	0.000	0.000	0.000	0.000	

**Correlation is significant at the 0.01 level (2-tailed)

9.3.3 Multiple Regression Analysis

In this step multiple regression analysis was employed to test the relationship between five predictors and sustainability awareness. From the Coefficient table (Table 9.4) it is observed that all five predictors had a tolerance value > 0.10 and their VIF (variance inflation factors) value ranges from 1.243 to 1.619 thus ensuring strongly that no multi collinearity exists. Further from the model summary, the coefficient of determination R2 was 0.867, indicating 86.7% sustainability awareness can be achieved because of five factors, and the R value was 0.931. Thus, the effect size of this research is found to be large. The multiple regression model was obtained which produced $R2 = 0.865$, $F = 504.100$, $p < 0.05$. This indicated that overall model is statistically significant at $p = 0.000 < 0.05$. The individual model variables indicated that the material impact ($\beta = 0.693$, $p < 0.05$), process optimization ($\beta = 0.269$, $p < 0.05$), economical aspects ($\beta = -0.060$, $p < 0.05$), social aspects ($\beta = 0.035$, $p < 0.05$) and key drivers ($\beta = 0.069$, $p < 0.05$) were statistically significant with sustainability awareness. However economic aspects have significant but negative relation with sustainability awareness. It is believed that while responding, respondents may have considered the fact that improvements at any stage will have their economic consequences without considering long-term benefits. Thus, it may have significance but has a negative relation. Following Regression, an equation was developed using Table 9.4, representing regression model to explain liner relation between sustainability awareness and its independent factors.

Table 9.2 Constructs and items mentioned in the questionnaire for survey

Factors	Indicators	
Sustainability	Awareness of sustainability concept	Influence on R & D practices
	Involvement of management	Increase transparency
	Improve supportive functions	Setting benchmark companies
	External environment	Impact on manufacturing cost
	Code of conduct for supplier	Improvement in org. performance
	Internal sustainability policy	R & D during design phase
	Internal motivation and external stimuli	Understanding of environ. aspect
	External pressure	Incorp. in vision and mission
	Impact on stakeholders	Envi. impact as funct. requirement
Environmental aspects	Material impact on environment	Consider biodegradable materials
	Identification of hot spots	Waste reduction (zero landfill)
	Energy consumption and CO_2 emission	Use of recyclable product
	Consideration of GHG emission	
Economic aspects	Purchase materials from certified supplier	Considering environmental aspect will improve financial performance
	Reduced energy and resource consump	Invest to minimize CO_2 emission
	Use of recyclable packaging material	
Social aspects	Consider health and safety issues	Provide opportunity for employees
	Incorp. of edu. and training of employees	Ensures employees participations
	Transparency in organizational policies	
Key drivers	Global initiative	Policies for green project
	Government initiatives	Use of LCA and other metrics
	Research and development facility	

Source Peukert and Sahr (2010), Mitra (2012), Watson (2012), Lozano et al. (2016), Chaturvedi et al. (2017)

$$SustainabilityAwareness = -0.075 + 0.693M.I. + 0.269P.O. \\ -0.060E.A. + 0.035S.A. + 0.069KD \quad (9.1)$$

In Eq. (9.1), *M.I.* = Material impact, *P.O.* = Process optimization, *E.A.* = Economic aspect, *S.A.* = Social aspect and *KD* = Key driver.

Therefore, hypothesis H_1 was supported (Tables 9.3 and 9.4).

9.4 Discussion

The results indicate existence of sustainability awareness in the Indian pharmaceutical industry. However, passé to adopt related practices is slow. This concurs with the findings of Goyal (2014), Kolk (2010) and Thijssens et al. (2016). Findings from this research support the hypothesis. The observed value of Mean i.e. 3.46–4.53 (Table 9.3) shows that organizations are accepting that sustainability awareness can be explored by considering practices pertaining to TBL and in presence of key drivers. Further beta value from regression equation shows significant amount of variance in the dependent variables i.e. sustainability awareness because of each independent variable. On the basis of findings, it is highlighted that sustainability awareness is mainly driven by consideration of material impact, process optimization, economic aspects, social aspects and presence of key drivers.

9.5 Implications

A major contribution of this paper is the statistics-based regression model, which integrates elements of TBL, key drivers and sustainability awareness. Contributions of findings are useful to explore the independent influences of predictors on sustainability awareness. This regression model is considered as a part of ongoing research thus it will be helpful to develop sustainability model whose use can be explored in the Indian pharmaceutical industry to achieve overall sustainability. Since regression model has been developed by direct survey method and statistical tools to analyze collected data have been used, it is thus deemed to be considered suitable for its practical usage. Since every firm has its own decision-making criteria thus managers have to fix up their priorities related to TBL and key drivers to enhance sustainability awareness by looking on hazardousness' and impact of used material, thus can take proactive actions.

9.6 Conclusions

Theoretical development and statistical analysis support this research work which builds up on the hypothesis that practices related to TBL and key drivers are useful and can be utilized as important factors to explore sustainability awareness in Indian pharmaceutical industry. More importantly, this research can possibly be used to diagnose

Table 9.3 Item statistics

Item	Mean	S.D.	N
1. Proper understanding of concept	4.39	0.750	393
2. Involvement of management	4.28	0.796	393
3. Incorporation of the concept in the vision and mission	4.09	0.830	393
4. Improvement in supportive function	3.81	0.854	393
5. External environment	3.70	0.731	393
6. Organization has code of conduct for suppliers	4.04	0.755	393
7. Organization has internal sustainability policy	4.16	0.882	393
8. Internal motivations	4.12	0.781	393
9. External pressure	3.46	1.070	393
10. Impacts on stakeholders	3.53	1.020	393
11. It influence research and development facilities	4.19	0.611	393
12. Transparency policy	4.21	0.840	393
13. Benchmark for other companies	4.09	0.714	393
14. Manufacturing cost	3.88	0.867	393
15. Overall performance	4.30	0.731	393
16. Through formal management	3.96	0.844	393
17. During product design phase	3.75	0.830	393
18. Environmental impacts as functional requirement	3.77	0.907	393
19. Considers impact of material used for mfg	4.04	0.706	393
20. By identifying hot spots of manufacturing process	3.72	0.901	393
21. Consider process specific energy and CO_2 emissions	3.58	0.865	393
22. Consider green house gas (GHG) emissions	3.77	0.926	393
23. Uses biodegradable materials	4.07	0.842	393
24. Ensures zero landfill (effluent and waste)	3.93	0.884	393
25. Use of recyclable material	3.98	0.973	393
26. Purchase materials from certified suppliers	4.09	0.689	393
27. Invest to minimize energy and resource consumption in the production	4.18	0.571	393
28. Use recyclable packaging materials	4.05	0.971	393
29. Consider financial implications in purchasing	3.86	0.743	393
30. Health and safety issues	4.30	0.706	393
31. Education and training of employees	4.30	0.755	393
32. Transparency in policies	4.18	0.782	393
33. Opportunities for employees	4.00	0.732	393
34. Employee's participation	3.86	0.789	393
35. Globally driven initiatives	3.98	0.612	393
36. Government initiative	3.88	0.734	393

(continued)

Table 9.3 (continued)

Item	Mean	S.D.	N
37. Research and development facility	4.21	0.750	393
38. Policies for green projects	4.18	0.759	393
39. Innovative thinking such as use of green chemistry, energy efficient manufacturing process, use of LCA of product etc.	3.93	0.863	393

Table 9.4 Coefficients used in the model

Model 1	Unstandardized coefficients	Standardized coefficients	t	Sig.	Collinearity statistics	VIF
	B	Beta			Tolerance	
(Constant)	−0.075		−0.732	0.465		
Material impact	0.693	0.749	34.41	0	0.726	1.377
Process optimization	0.269	0.288	13.29	0	0.732	1.365
Economic aspects	−0.06	−0.072	−3.05	0.002	0.618	1.619
Social aspects	0.035	0.049	2.38	0.018	0.804	1.243
Key drivers	0.069	0.073	3.22	0.001	0.661	1.513

and analyze which practices of TBL and key drivers are effective in influencing more on exploring sustainability awareness. The findings of this work indicate that consideration of material impacts and related process optimizations have strong association with organizational environmental performance thus leading to enhanced sustainability awareness. Conversely, managers have to accept the challenges of economic consequences associated with immediate investments and long-term return with sustainability awareness in the Indian pharmaceutical industry. In addition to these, this research provides a broad understanding of how different aspects of TBL and key drivers can be adopted as an informative tool to explore sustainability awareness, thus to achieve overall sustainability in Indian pharmaceutical industry.

References

Agar AG, Arcese G, Lucchetti MC (2016) Waste management and environmental impact: a case study of pharmaceutical industry. Pathways Environ Sustain 97–106 (2016). https://doi.org/10.1007/978-3-319-03826-1_10

Amran A, Ooi SK (2014) Sustainability reporting: meeting stakeholder demands. Strat Direct 30(7):38–41. https://doi.org/10.1108/SD-03-2014-0035

Cespi D, Beach ES, Swarr TE, Passarini F, Vassura I, Dunn PJ, Anastas PT (2015) Life cycle inventory improvement in the pharmaceutical sector: assessment of the sustainability combining PMI and LCA tools. Green Chem 17:3390–3400. https://doi.org/10.1039/c5gc00424a

Chaturvedi U, Sharma M, Dangayach GS, Sarkar P (2017) Evolution and adoption of sustainable practices in the pharmaceutical industry: an overview with an Indian perspective. J Clean Prod 168:1358–1369. https://doi.org/10.1016/j.jclepro.2017.08.184

Gopal PRC, Thakkar J (2015) Sustainable supply chain practices: an empirical investigation on Indian automobile industry. Prod Plan Control. https://doi.org/10.1080/09537287.2015.1060368

Goyal N (2014) Corporate sustainability reporting practices among Indian companies—myth or reality. Int J Manage Social Sci Res 3(1):54–60. ISSN: 2319-4421

Hair JF Jr, Black WC, Babin BJ, Anderson RE, Tatham RL (2006) Multivariate data analysis, 6th edn. Pearson Education, Upper Saddle River, NJ

Kolk A (2010) Trajectories of sustainability reporting by MNC's. J World Bus 45(4):367–374. https://doi.org/10.1016/j.jwb.2009.08.001

Lozano R, Nummert B, Ceulemans K (2016) Elucidating the relationship between sustainability reporting and organisational change management for sustainability. J Clean Prod 125:168–188. https://doi.org/10.1016/j.jclepro.2014.03.031

Malhotra NK (2004) Marketing research—an applied orientation, vol 4. Pearson, Prentice Hall, Upper Saddle River, NJ

Mitra PK (2012) Sustainability reporting practices in India: its problem and prospects. Int J Market Finan Serv Manage Res 1(5):109–115

Nunnally JC (1978) Psychometric methods. Mc-Graw Hill, NY

Peukert J, Sahr K (2010) Sustainability in the chemical and pharmaceutical industry—results of a benchmark analysis. J Bus Chem 10(7):97–106

Raman RS (2006) Corporate social reporting in India, a view from the top. Global Busin Rev 7(2):314–324 (2006). https://doi.org/10.1177/097215090600700208

Sheldon R (1993) A: chirotechnology: industrial synthesis of optically active compounds. Markcel Dekker, New York, pp 47–61

Sheldon R (2016) A: the E factor 25 years on: the rise of green chemistry and sustainability. Green Chem 19(1):18–43. https://doi.org/10.1039/C6GC02157C

Sihvonen S, Partanen J (2016) Implementing environmental considerations within product development practices: a survey on employees' perspectives. J Clean Prod 125:189–203. https://doi.org/10.1016/j.jclepro.2016.03.023

Thijssens T, Bollen L, Hassink H (2016) Managing sustainability reporting: many ways to publish exemplary reports. J Clean Prod 136:86–101. https://doi.org/10.1016/j.jclepro.2016.01.098

United Nations Environmental Programme (UNEP) report. http://web.unep.org/. Last accessed 20 May 2019

Watson WJW (2012) How do the fine chemical, pharmaceutical, and related industries approach green chemistry and sustainability. Green Chem 14(2):251–259 (2012). https://doi.org/10.1039/c1gc15904f

Wernet G, Conradt S, Isenring HP, Gonzalez CJ, Hungerbuhler K (2010) Life cycle assessment of fine chemical production: a case study of pharmaceutical synthesis. Int J Life Cycle Assess 15(3):294–303. https://doi.org/10.1007/s11367-010-0151-z

10

Artificial Neural Networks Techniques for Wind Speed Prediction

Sumanta Pasari, Aditya Shah, and Utkarsh Sirpurkar

Abstract Renewable energy sources are one of the most vital alternatives to the conventional non-replenishable energy generating systems. Among several renewable power sources, the installed wind power capacity contributes to almost half of the total capacity. However, the variability and seasonality in wind speed, wind direction, atmospheric pressure, relative humidity and precipitation cause wind power generation to be highly volatile. In this regard, the present study aims to develop a wind speed prediction scheme using artificial neural network (ANN) techniques. Single step and multistep recurrent neural networks (RNNs) are implemented. The long short term memory (LSTM), rectified linear unit (ReLU) activation function and Adam optimization algorithm are considered to carry out daily to monthly prediction using the RNN process. Results, based on the data from Charanka solar energy park in Gujarat, indicate that root means square (RMS) errors for univariate single layer, multivariate single layer and univariate two-layer models are 0.601, 0.782 and 1.120, respectively. Therefore, a univariate single layer RNN architecture is recommended for wind speed prediction. We envisage that a multilayer RNN model may improve the prediction accuracy over longer period.

Keywords Wind energy prediction · Neural network · Charanka solar park

---

S. Pasari (✉) · A. Shah · U. Sirpurkar
Department of Mathematics, Birla Institute of Technology and Science, Pilani Campus, Pilani 333031, RJ, India
e-mail: sumanta.pasari@pilani.bits-pilani.ac.in

A. Shah
e-mail: f2014715p@alumni.bits-pilani.ac.in

U. Sirpurkar
e-mail: f2014785p@alumni.bits-pilani.ac.in

10.1 Introduction

The non-renewable fossil fuels, which include coal, oil and natural gas, play a dominant role in global energy consumption and industrial revolution. Currently fossil energy supplies to about 80% of the total worlds' energy. However, these carbon-rich reserved fossil fuels are not only exhaustible in about next ten decades, but also the primary source of air pollution, carbon emission and greenhouse gas production (Bhatia and Gupta 2018; Kumar et al. 2010; Wang et al. 2011). The planners and policy makers must therefore look for alternative energy sources to meet the increasing energy demand of developing economies in a sustainable way.

Renewable energy sources such as wind, solar radiation, hydroelectricity and biomass are regenerative and abundant. The cost of renewable power production, after a decade, will be much lower than the energy production cost from fossil fuels (Kumar et al. 2010). In India, as of 2018, about 20% of the total installed power comes from these clean energy resources against the national target of 40% by 2030. India sets for an ambitious goal of generating 100 GW of solar and 60 GW of wind power by 2022, and 250 GW of solar and 100 GW of wind power by 2030 (Bhatia and Gupta 2018). Achieving such an aspiring target requires dedicated research in various fields including technology development, site-identification, smart-grid planning, energy integration (to the main power grids), policy making, cost-benefit analysis and energy forecasting (Bhatia and Gupta 2018; Kumar et al. 2010; Wang et al. 2011; Rather 2018). In this study, we concentrate on wind energy prediction based on the artificial neural network (ANN) techniques.

Wind energy, like solar power, is plentiful, replenishable and a promising alternative to burning fossil fuels. It uses air flow through turbines to provide mechanical power convertible to electricity. Neither water is consumed during operation, nor is more land required. Wind farms comprise many individual onshore or offshore wind turbines connected to the electric power transmission network. However, wind power supply is highly volatile due to the variability and seasonality in heat energy from the sun, wind speed, wind direction, atmospheric pressure, relative humidity, precipitation and temperature gradients between land and sea (Kumar et al. 2010). Improved wind power predictions thus are crucial in effective market design, real-time grid management, power transmission capacity evaluation and ancillary information (Wang et al. 2011).

There have been several attempts to wind energy prediction, varying from deterministic (physical) approach based on numerical weather prediction to statistical technique based on historical data analysis, time-series modeling or artificial intelligence (neural networks) (Bhatia and Gupta 2018; Kumar et al. 2010; Wang et al. 2011; Rather 2018). On the basis of different time-horizons, wind power prediction could be classified as immediate-short-term (up to 8 h), short-term (day ahead) and long-term (multiple days to months ahead) prediction (Wang et al. 2011). While real-time grid operations, energy trading and regulatory actions benefit from immediate-short-term prediction, economic load balance and operational security analysis depend on the short-term wind energy prediction. The long-term forecasting is useful for cost-optimal energy storage strategy, operation management, maintenance planning and cost-benefit analysis (Bhatia and Gupta 2018; Kumar et al. 2010; Wang et al. 2011). However, due to the volatility of wind speed over the months, it is challenging to accurately predict the wind speed for a number of practical applications. Therefore to improve the short/long term forecasting accuracy, several methods have been proposed such as direct multistep recurrent neural network (RNN) using LSTM neural network combining fuzzy entropy (Qin et al. 2019), multivariable (e.g., wind speed, temperature, humidity and pressure) stacked LSTMs model (MSLSTM) (Liang et al. 2018), univariate and multivariate autoregressive integrated moving average (ARIMA) with ANNs (Cao et al. 2012) and pipelined recurrent neural network (PRNN) based NARMAX ANN model (nonlinear auto regressive moving average artificial neural network with external inputs) (Liangyou et al. 2019). Nonetheless, there is no global best prediction model to be applicable irrespective of geographical region (Rather 2018; Qin et al. 2019; Liang et al. 2018; Cao et al. 2012; Liangyou et al. 2019). Thus, in the present work, we focus on daily to monthly wind speed prediction using a data driven single step and multistep recurrent neural network (RNN) process for the Charanka solar park of India.

10.2 Data Description

Our dataset comprises hourly wind speed and wind direction data of Charanka solar park (also known as Gujarat solar park 1; 23.95° N, 71.15° E) for a period of 15 years, 2000–2014. This dataset is publicly available from the National Solar Radiation Database maintained by National Renewable Energy Laboratory (NREL) (NREL homepage 2019). In addition to wind speed and wind direction, the dataset also provides information of temperature, solar radiation, such as DHI (Diffuse Horizontal Irradiance), DNI (Direct Normal Irradiance), GHI (Global Horizontal Irradiance), solar zenith angle, clear-sky DHI, clear-sky DNI and clear-sky GHI along with the information of some meteorological parameters such as dew point, atmospheric pressure, relative humidity, precipitable water and snow depth.

10.3 Methodology

The methodology for the present study comprises three major steps: data preparation, RNN model implementation and wind speed prediction. The RNN model implementation further consists of three sequential steps, namely training, validation and testing. Before we move on, some preliminary discussion about RNN process would be helpful.

RNNs are a type of artificial neural network with loops in them to persist information in sequences of inputs. The chain-like architecture in RNNs allow them not only to learn from training (similar to feed-forward neural networks), but also to remember things learnt from all prior inputs while generating outputs. The beauty and strength of RNN therefore comes from its *hidden state* which manages to span many time steps as it marches forward through sequences of input, output or both vectors (Graves 2012; An Introduction to Recurrent Neural Networks 2019). In order to describe the workflow for the RNN process at time step t, let x_t be the input vector, h_t be the hidden state and y_t be the output vector. Needless to mention that the output vector in RNN is influenced not only by the immediate current input, but also the entire history of inputs (Graves 2012; An Introduction to Recurrent Neural Networks 2019). The hidden state, in general, can be mathematically expressed as

$$h_t = f(h_{t-1}, x_t) \tag{10.1}$$

The hidden state h_t is updated as

$$h_t = f(W_{hh}h_{t-1} + W_{xh}x_t) \tag{10.2}$$

Thus h_t is a function of the input x_t modified by a weight matrix W_{xh} (as used in feed-forward networks) added to the previous hidden state h_{t-1} multiplied by the transition matrix (hidden state to hidden state matrix) W_{hh}. These two weight matrices serve as filters by providing appropriate weights (importance) to the present input and past hidden state. The sum of these two intermediates, weighted input and hidden state, is then squashed by the function f, either a logistic sigmoid, tan-hyperbolic, ReLU (rectified linear unit) or others (Graves 2012; An Introduction to Recurrent Neural Networks 2019). Using the hidden state h_t, modified by the weight matrix W_{hy}, the output vector y_t is computed as

$$y_t = W_{hy}\, h_t \tag{10.3}$$

Comparing the model output to the actual (target) output, the error is generated. This error is then back-propagated to update the weights recursively until an allowable error limit is achieved. In this way, the RNN gets trained via the long short term memory (LSTM) technique (Graves 2012). As RNN has a memory and can remember every information through time, it is a powerful tool for time series prediction (Graves 2012; An Introduction to Recurrent Neural Networks 2019).

In this work, we implement three different RNN architectures, namely univariate unit-step single layer, multivariate two-step single layer and univariate multistep two-layer model for one-day to one-month wind speed prediction. While the first model considers single step inputs, that is, inputs from only the immediate previous state to predict future values, the second model considers previous two states for prediction. The multistep model uses inputs from several earlier steps for providing outputs. For each model, Adam optimization algorithm is used to update network weights iteratively based on training data, whereas the loss function is evaluated on the basis of mean absolute error (MAE). These supervised algorithms utilize about 70% of total data points for training and remaining 30%, divided equally, for model

Table 10.1 Parameters used in different RNN models

	Univariate unit-step single layer	Multivariate two-step single layer	Univariate multistep two-layer
Hidden layers	1	1	2
Input neurons	1	15	1
Steps	1	2	30
Neurons in hidden layer	100	100	100
Epochs	10000	200	100
Learning rate	0.001	0.001	0.001
Activation function	tan-hyperbolic	ReLU	ReLU
Optimization	Adam	Adam	Adam
Loss function	MAE	MAE	MAE

validation and testing. A comparison of the parameters used in these three different models is provided in Table 10.1.

For the univariate unit-step single layer RNN model, entire 15 years of daily wind speed data is used for training towards next day's wind speed prediction. Using this one-day predicted information, model then predicts the wind speed of the day next to the previously forecasted day. The process continues to output 365 days (one day at a time) wind speed prediction. In the multivariate two-step single layer model, daily wind speed data along with other 14 variables, such as wind direction, temperature, solar zenith angle, DHI, DNI, GHI, clear sky DHI, clear sky DNI, clear sky GHI, dew point, pressure, relative humidity, precipitable water and snow depth is considered. Recall that, physically, the wind speed depends on many variables like solar radiation and other meteorological parameters. This multivariate RNN also considers previous two steps while forecasting future wind speed data. The univariate multistep two-layer model, in contrary to the previous ones, uses two LSTM layers to predict one month's wind speed data with an input of current month's wind speed data.

The entire methodology is implemented in python using open-source *keras* library with *TensorFlow* backbend. Experimental results are provided in the next section.

10.4 Results and Conclusions

The root means square (RMS) errors corresponding to univariate unit-step single layer, multivariate two-step single layer and univariate multistep two-layer model for the year 2014 turn out to be 0.601, 0.782 and 1.120, respectively. The predicted wind speed is plotted against the actual wind speed in Fig. 10.1. It clearly demonstrates that complex architectures in multivariate RNN or univariate multistep RNN does not produce desirable fit. The reason for such poor performance could be due to the

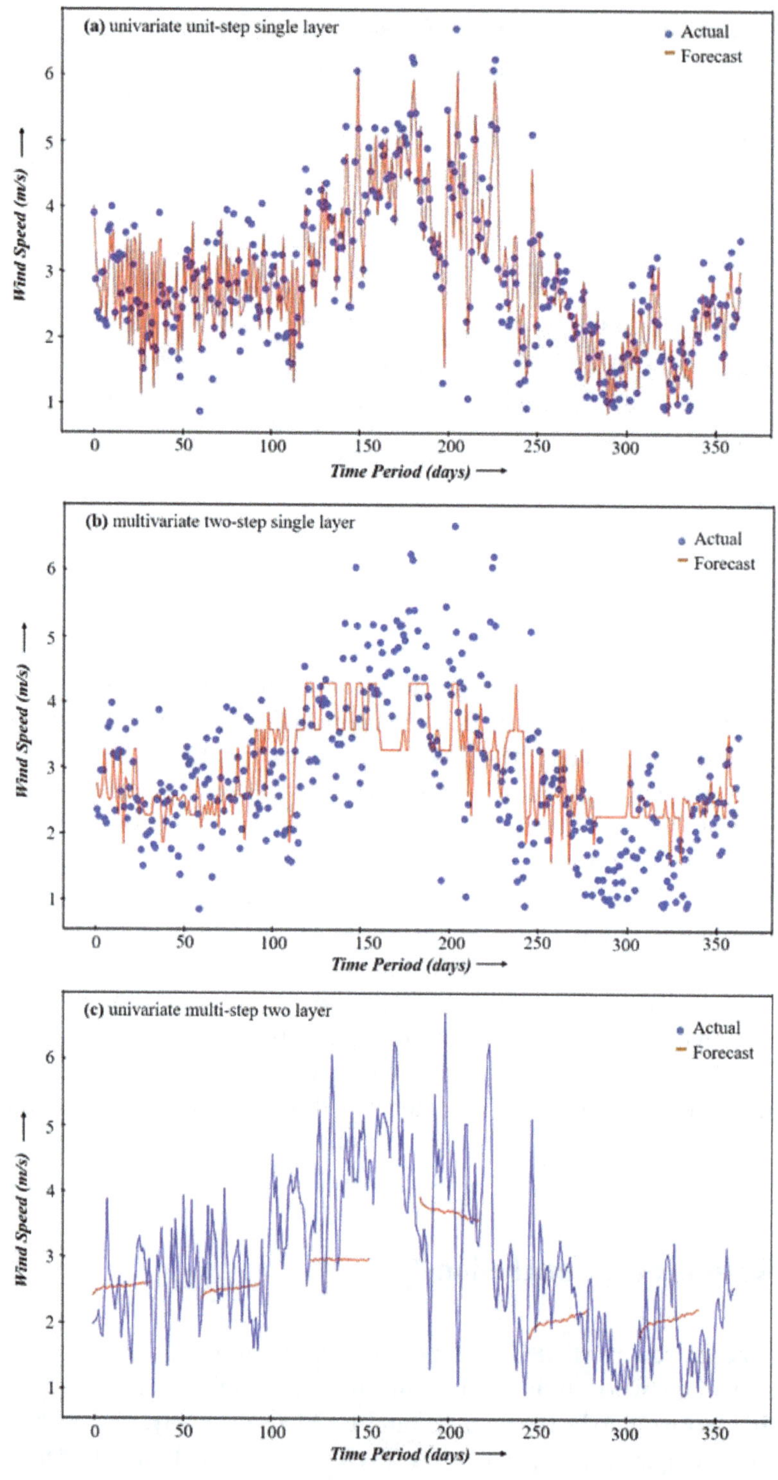

Fig. 10.1 Predicted versus actual wind speed corresponding to **a** univariate unit-step single layer, **b** multivariate two-step single layer and **c** univariate multistep two-layer RNN architecture (forecasts of six alternative months are shown)

inclusion of several variables in multivariate RNN or the incorrect choice of training method and/or activation function.

Thus, the present study leads to the conclusion that a simple univariate unit-step single layer RNN model is the most suitable architecture for short-term wind speed prediction. The proposed scheme is equally applicable to forecast other renewable energy sources, such as solar radiation.

References

An Introduction to Recurrent Neural Networks (2019). https://medium.com/explore-artificial-intelligence/an-introduction-to-recurrent-neural-networks-72c97bf0912, Accessed 24 June 2019

Bhatia SC, Gupta RK (2018) Textbook of renewable energy. Woodhead Publishing India, pp 1–334

Cao Q, Ewing BT, Thompson MA (2012) Forecasting wind speed with recurrent neural networks. Eur. J Oper Res 221(1):148–154

Graves A (2012) Supervised sequence labelling with recurrent neural networks. Springer, Berlin, pp 1–146

Kumar A, Kumar K, Kaushik N, Sharma S, Mishra S (2010) Renewable energy in India: current status and future potentials. Renew Sustain Energy Rev 14:2434–2442

Liangyou H, Dongxiang J, Qian H, Yongshan D (2019) Wind speed forecasting using fully recurrent neural network in wind power plants. https://pdfs.semanticscholar.org/fef5/4b2b2ae713ebf8b08ad963d7dbbea15ad2fc.pdf, Accessed 24 Aug 2019

Liang S, Nguyen L, Jin F (2018) A multivariable stacked LSTM network for wind speed forecasting. In: 2018 IEEE conference on big data. https://arxiv.org/pdf/1811.09735.pdf, Accessed 24 Aug 2019

NREL homepage. https://www.nrel.gov/, Accessed 24 June 2019

Qin Q, Lai X, Zou J (2019) Direct multistep wind speed forecasting using LSTM neural network combining EEMD and fuzzy entropy. Appl Sci 9(126):1–19

Rather NUR (2018) Introduction to renewable energy technologies in India. Education Publishing, pp 1–140

Wang X, Guo P, Huang X (2011) A review of wind power forecasting models. Energy Procedia 12:770–778

11

Electro-Pneumatic Handling Systems and their Monitoring through Data Analytics

Maximilian Rolinck, Sebastian Gellrich, Christoph Herrmann, and Sebastian Thiede

Abstract Ensuring constant high process and product quality is one of the decisive competitive factors in today's industrial production. Industry 4.0 by means of the analysis of production data is seen as a viable approach to achieve zero defects production. Due to their sensitive behavior, energy data, which is often acquired independently of the machine control, can provide an adequate data source for quality-related analysis tasks. An analysis based on energy data reduces the effort for accessing machine controls and enables scalability. This paper shows the concept and implementation of a data-driven process and quality monitoring tool based on energy data for an electro-pneumatic handling system.

Keywords Cyber Physical Production System · Data Mining · Process and quality monitoring

11.1 Introduction

An effective quality control is one of the key challenges in modern production. In order to foster a data-driven quality control, monitoring systems serve as a viable approach. In highly automated production industries electro-pneumatic automation systems gain importance (Statista 2019) and a higher spread in several industries that are sensitive due to hygienic requirements as well as explosive materials (Watter 2017). For this reason, this paper shows an implementation of a data-driven quality control tool for electro-pneumatic automation systems in terms of a Cyber-Physical Production System (CPPS). The CPPS takes the energy data of the electro-pneumatic automation system as input for process and quality monitoring, due to scalability and independence from the machine control. Therefore, background information on

M. Rolinck (✉) · S. Gellrich · C. Herrmann · S. Thiede
Chair of Sustainable Manufacturing and Life Cycle Engineering, Institute of Machine Tools and Production Technology (IWF), Technische Universität Braunschweig, Langer Kamp 19b, 38106 Braunschweig, Germany
e-mail: m.rolinck@tu-braunschweig.de

CPPS is briefly introduced in Sect. 11.2. On this basis, the concept of a process and quality monitoring tool for an electro-pneumatic handling system is developed in Sect. 11.3. The concept is exemplified within the experimental set-up of the Joint Indo-German Experience Lab at BITS Pilani, India.

11.2 Research Background

Facilitated by technological improvements, Industry 4.0 and digital manufacturing are major trends in industry and research (Lu 2017; Vaidya et al. 2018). CPPS are proposed to be a promising approach to pursue these trends as well as future demands in production (Thiede et al. 2014). According to Thiede et al. CPPS are based on the "implementation of elements from cyber physical systems" which contain "computational and physical capabilities combined with the possibility of human machine interaction" (Thiede et al. 2014). For the assessment and implementation of a CPPS Thiede et al. give an appropriate framework. The framework describes a loop of different key aspects and ranks them in three levels each, as shown in Fig. 11.1. In general, the framework describes a continuous data and information processing between the physical world and the so-called cyber world, which are connected by the elements data acquisition and feedback/control.

To perform analysis in an automated model, data mining gives various possibilities to process data. It can operate six different tasks according to Larose: *Description, estimation, prediction, classification, clustering* and *association*. These tasks depend on the intended analysis target and require different data pre-preparation steps (Larose and Larose 2014).

In order to fulfil the introduced six tasks, several methods are available. Different data mining methods perform one or more of the previously named tasks (Chakrabarti

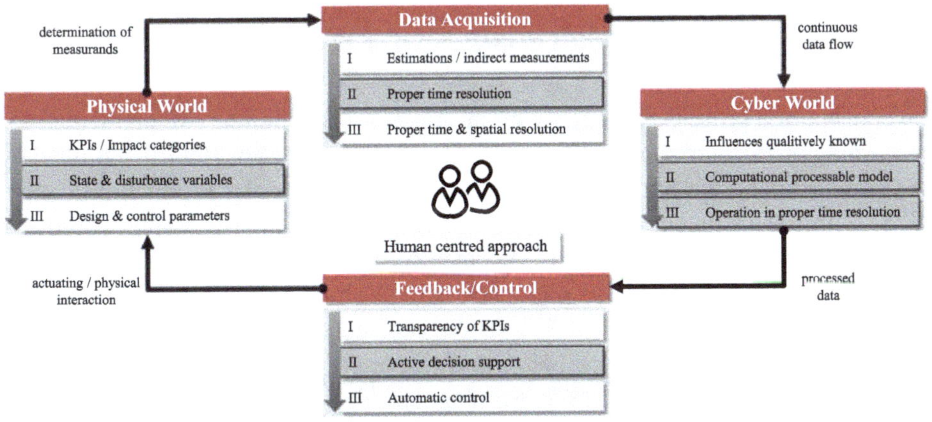

Fig. 11.1 CPPS according to Thiede et al. (2014) with levels highlighted in grey for the concept to be assigned (see Sect. 11.3). Data flows between elements are depicted by black arrows

et al. 2009). Besides that, the methods can be differentiated into supervised and unsupervised learning. The so called supervised learning is used for estimation, prediction and classification and includes the allocation of target variables to predictor variables. This allocation is also known as labelling. Unsupervised learning is used for clustering, association as well as description and exposes patterns out of unlabeled datasets (Chakrabarti et al. 2009). In order to deploy, for example, a classification task, a supervised learning method that performs a categorical statement based on a labelled dataset is chosen.

11.3 CPPS-Based Process and Quality Monitoring Tool

11.3.1 Concept

For the conceptual design of a process and quality monitoring tool, the framework of the introduced CPPS is used. In order to provide advanced support to an operator of a machine or production system, accordance with a high level (see Fig. 11.1) in each element is vital. Thus, *state variables* (II) from the physical world are chosen and gathered at least in *proper time resolution* (II) in a form of data acquisition. In order to ensure the independence of further data processing from predetermined software solutions or adjustments of machine controls for data access, data gateways can be deployed. Implementing a *computational processable model* (II) in the cyber world and processing data in *proper time resolution* (III) results in a higher informational degree about the current state of the machine or production system. The feedback of this information in form of *active decision support* (II), e.g. alerts or instructions, enables the operator to effectively interact with the physical world.

11.3.2 Exemplary Implementation

The concept of a process and quality monitoring tool is exemplary implemented for the learning platform of the Joint Indo-German Experience Lab (JInGEL), funded by the German Academic Exchange Service (DAAD). A Modular Production System (MPS) within the small-scaled production line simulating an assembling process is chosen for this purpose, called pick & place (p&p) process in the following. The central task of this station is to pick caps from a chute and to place them on the provided workpieces (see Fig. 11.2, left). Workpieces are distributed to the station one after another on a conveyor belt and transported to the middle of the station. Each piece is hold back by an electro-magnetic driven deflector and meanwhile detected by a sensor. When a workpiece is fixed, the p&p process starts. A suction cup, mounted on an electro-pneumatic driven handling arm, picks one cap from the chute. After being elevated, it is placed on the workpiece. The assembled product is

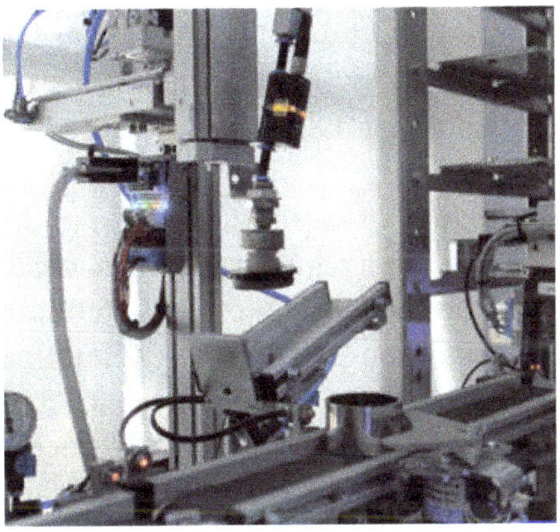

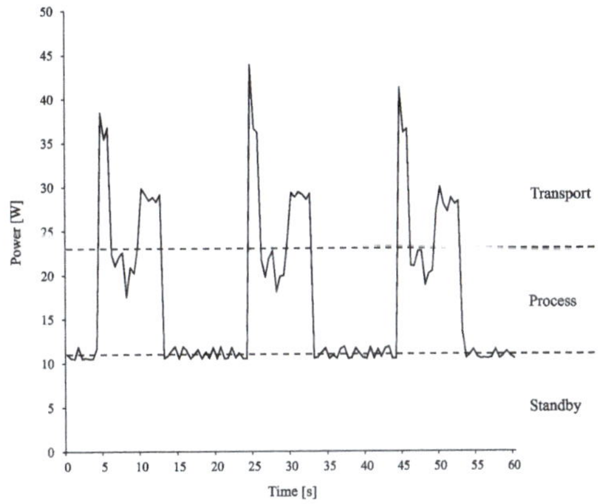

Fig. 11.2 *Left* p&p process within the experience lab at BITS Pilani – a cap is elevated from the chute; *right* classes of power demand of the three p&p process states

released and transported to the next station. In case that the chute runs empty and no cap is available for the assembling process, the handling arm stops the suction process after detection, but the unassembled workpiece is still transported to the next station. This results in a considerable lack of product quality that should be detected by the monitoring tool. The product is sorted out in one of the following production steps, but the p&p process runs incorrectly until the operator notices the error and intervenes. In any case, the process runs at a fixed time interval of 20 s. This time stamp serves as a basis for further modeling.

Physical World and Data Acquisition. During the process, the energy state variables current and electrical power as well as compressed air flow are measured. Using a coupler from Beckhoff, the data is converted into E-bus signal representations and transported by an EtherCAT® network. Via the Visual Studio® embedded TwinCAT® software an OPC UA server is installed, which represents a data gateway, enabling data to be accessible for further processing. In order to process data, the flow-based development tool Node-RED is chosen. It provides a platform for functions programming and integration of versatile input and output options. The provided data is stored in a MySQL database which facilitates the computation of the process's energy demand from the measured electrical power. Similar data processing is applied for the compressed air flow giving the average consumption of compressed air. These five state variables enable further analyses that are depicted in the following section. Besides these variables, control parameters are given due to the possibility for an operator to intervene the process by replacing caps or even stopping the process.

Cyber World. For data exploration, data is extracted in a time resolution of 0.5 s. Examining the results for the power demand over three process cycles (Fig. 11.2, right), three different states can be identified. The peaks in the curve classify the demand when a workpiece is transported by the conveyor belt. The transport state is defined for the maximum electrical power demand at the time the conveyor belt is running. The process state is identified by the distinct p&p process itself when the conveyor belt is not running, the workpiece is fixed by the deflector and the electro-pneumatic driven handling arm is manipulating the product. The third state represents the standby mode when a base load at approximately 11 W is caused by the machine control and lighting.

The process is divided into two different conditions of cap availability, i.e. the quality criteria of the p&p process. Condition 1 is used for a regular process in which caps are available. A process in which caps are missing and the workpiece is therefore not properly assembled is further referred to as condition 2. The data exploration shows, that no significant difference can be found between both cases in the level of power demand. However, a difference in the time sequence and duration of the process can be observed. In condition 2, the process runs two and a half seconds faster as the handling arms movement is aborted. Regarding the compressed air flow, condition 2 differs from condition 1 with a lower flow rate of about 5 l/min. Consequently, it is determined that condition 1 and condition 2 can be clearly distinguished in their energy consumption and average compressed air flow. The values for energy intensity, which is the energy demand for one process cycle of 20 s, and average compressed air flow over the same interval are depicted in the scatter plot in Fig. 11.3. Clear limits can be found at 350.4 Ws for the electrical energy intensity and 7.95 l/min for the average compressed air flow, with condition 1 always showing a higher value than condition 2.

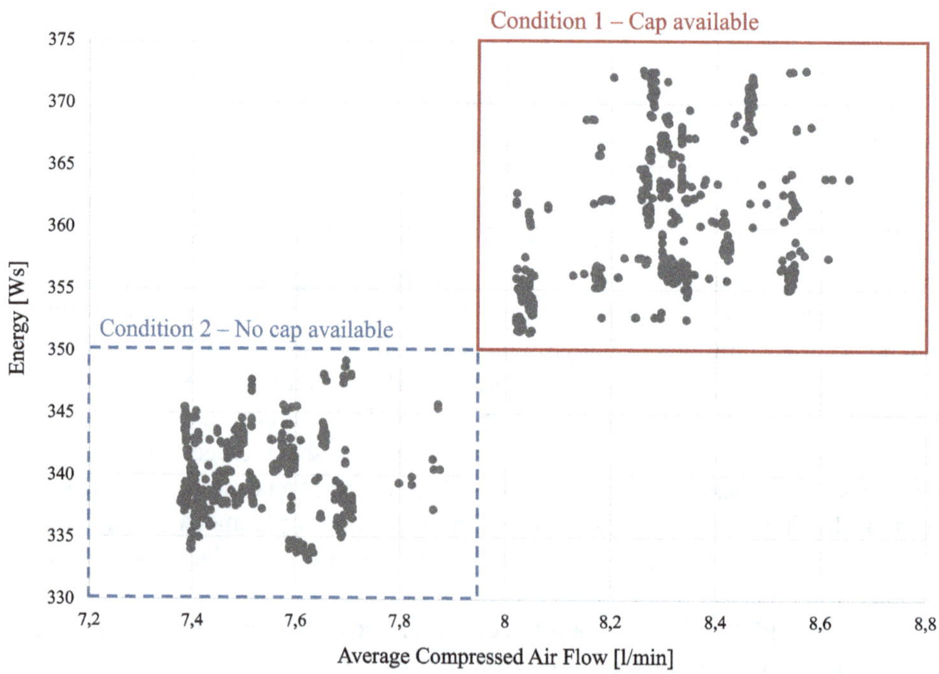

Fig. 11.3 Scatter plot on energy intensity and average compressed air flow for p&p process. Areas for condition 1 and 2 indicating quality criteria are marked

Table 11.1 Labelling data arrays for data preparation

Power (W)	Energy (Ws)	Current (A)	Compr. air flow (l/min)	Average air flow (l/min)	State	Condition
10.93	370.68	0.09	6.52	8.46	Standby	Cap
38.43	371.38	0.22	6.50	8.46	Transport	Cap
21.56	368.19	0.15	22.03	8.55	Process	Cap
33.71	337.08	0.15	17.23	7.39	Process	No cap
28.43	337.98	0.18	6.46	7.39	Transport	No cap
10.65	338.03	0.09	6.58	7.39	Standby	No Cap

These findings allow the classification of state variables regarding the state as well as the corresponding quality condition as shown in Table 11.1. For new real-time data, the classification is to be performed through a data mining approach based on supervised learning. In order to train a data mining method, a training data set with over 3,000 arrays of state variables is labelled according to the example shown in Table 11.1. Three supervised learning methods, i.e. random forest, support vector machines and decision trees, are examined. With an overall accuracy of 97.6% and good results in the F1-Score for the critical distinction of quality condition a decision tree based method is chosen for this use case. The trained model delivers information about the state (transport, process or standby) and the condition (with or without cap) of the process. In a next step, this information shall be provided to the operator via a monitoring tool.

Monitoring and Decision Support. In order to provide the operator with information, a dashboard is designed in Node-RED (Fig. 11.4). Information of two kinds are displayed, on the one hand real-time data about power demand and compressed air flow and on the other hand classifications on the current state and condition of the p&p system. The classifications allow the assignment of shares of energy intensity due to compressed air flow and electrical energy consumption per cycle and state. This information can be used for a dynamic Energy Value Stream Mapping, as shown in Fig. 11.4. In addition, direct feedback is given to the operator in case of a caps' absence. It appears in the form of a pop-up window which disappears only after an "Ok" button is confirmed and also instructs to replace caps at the p&p station. The implementation thus provides the operator with a process and quality monitoring tool. In a next step, key performance indicators like an Overall Equipment Effectiveness (OEE) could be implemented due to necessary parameters already being calculated by the system. In the current visualisation the amount of correctly assembled products over the last two minutes is depicted in a bar chart. Further sensor data could enrich this condition monitoring approach.

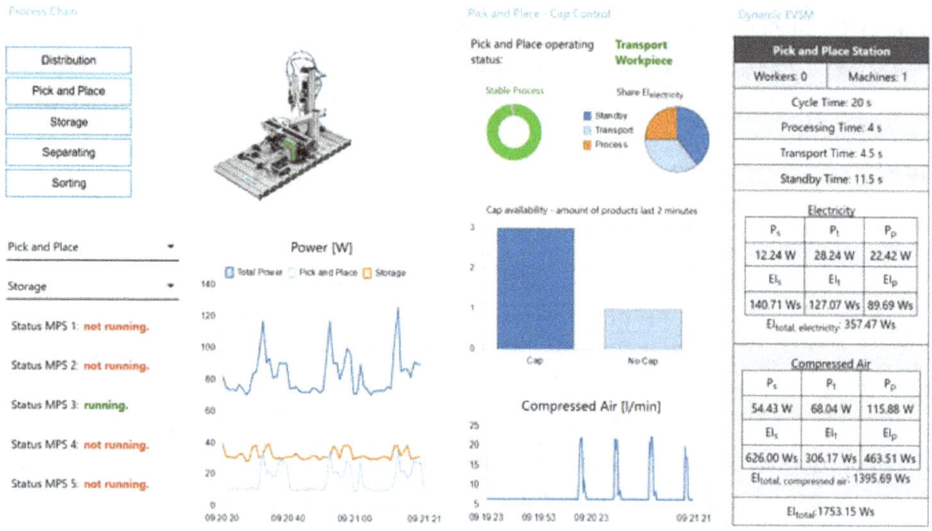

Fig. 11.4 Dashboard of the process and quality monitoring tool including real time data visualisation and data classification showing status report and dynamic Energy Value Stream Mapping

11.4 Conclusion and Outlook

In this paper it is shown how a process and quality monitoring tool on the basis of the CPPS framework can be implemented. By using energy data for monitoring of electro-pneumatic automation systems, independence from machine control and scalability can be guaranteed. The exemplary implementation shows that analytics of energy data can be a good way to achieve advanced support for an operator to run an electro-pneumatic production system. By providing an independent data gateway, further processing is not limited. The adaptation of the concept is thus possible also in industrial set-ups.

Acknowledgements The research for this paper is part of the project "JInGEL—Joint Indo German Experience Lab", a joint project between Technische Universität Braunschweig and Birla Institute of Technology and Science Pilani, and is funded by the German Academic Exchange Service (DAAD) under grant number 57219215. The authors are thankful for the funding and the support.

References

Chakrabarti S et al (2009) Data mining: know it all, 1st edn. Morgan Kaufmann, Elsevier, Amsterdam

Larose D, Larose C (2014) Discovering knowledge in data, 2nd edn. Wiley, Hoboken, New Jersey. https://doi.org/10.1002/9781118874059

Lu Y (2017) Industry 4.0: a survey on technologies, applications and open research issues. J Ind Inf Integr. 6:1–10. Elsevier, Amsterdam. https://doi.org/10.1016/j.jii.2017.04.005

Statista: Umsatz der Festo AG & Co. KG in den Jahren 2010 bis 2018. https://de.statista.com/statistik/daten/studie/444500/umfrage/umsatz-von-festo/. Last accessed 21 May 2019

Thiede S, Jurascheck M, Herrmann C (2014) Implementing cyber-physical production systems in learning factories. In: 6th CIRP conference on learning factories, pp 7–12. Elsevier, Amsterdam. https://doi.org/10.1016/j.procir.2016.04.098

Vaidya S, Ambad P, Bhosle S (2018) Industry 4.0—a glimpse. In: Procedia manufacturing, vol 20, pp 233–238. Elsevier, Amsterdam. https://doi.org/10.1016/j.promfg.2018.02.034

Watter H (2017) Hydraulik und Pneumatik, 5th edn. Springer Vieweg, Wiesbaden. https://doi.org/10.1007/978-3-658-18555-8

12

Electric-Load Intelligence (E-LI) systems: A Detailed Study

Nitesh Sihag and Kuldip Singh Sangwan

Abstract Energy and resource efficient manufacturing has become a key priority due to higher energy cost, market competition and environmental regulations. Better transparency and higher levels of disaggregation of energy data are necessary for energy efficiency improvement of machine tools. Since the beginning of the 21st century, some attempts have been made by the researchers to quantify the energy data but only up to the operational state of the machine tool. Better accuracy and transparency require disaggregation up to the component level. This study proposes an Electric-Load Intelligence (E-LI) system for identification of machine tool operating state and disaggregation of time and energy consumed up to the component level. The energy profile is obtained at the power input of a machine tool and analyzed using a set of signal processing techniques and load-disaggregation algorithms. The proposed methodology is validated through a case study of milling process. Various classifiers used in the disaggregation algorithms are compared for their accuracies using the case study data. The results reveal that only a small portion of the total cutting energy (782.24 kJ) was used for actual material removal (40.73 kJ). The proposed study provides accurate data in user friendly format to assist designers and manufacturers for strategic and economic decision making.

Keywords Electric-Load intelligence · Non-intrusive load monitoring · Energy disaggregation

---

N. Sihag · K. S. Sangwan (✉)
Department of Mechanical Engineering, Birla Institute of Technology and Science, Pilani-Campus, Pilani 333031, RJ, India
e-mail: kss@pilani.bits-pilani.ac.in

N. Sihag
e-mail: P2014009@pilani.bits-pilani.ac.in

12.1 Introduction

Energy efficiency of machine tools has become an important aspect of energy saving studies due to high energy consumption and low efficiency of machine tools. Also, electric energy consumption by machine tools is responsible for more than 99% of their carbon emissions (Li and Kara 2011). Hence, any reduction in energy consumption by machine tools will lead to economic as well as environmental benefits. During the early 21st century, the crude oil prices experienced an exponential surge from approximately 25 USD per barrel in 2001 to approximately 100 USD per barrel in 2008. This heightened the need for energy saving; and energy consumption emerged as an important criterion for modeling and optimization studies (Camposeco-Negrete 2015; Kant and Sangwan 2014; Zhang et al. 2018). The research for the improvement of surface finish and tool life, and reduction of cutting forces by optimizing cutting parameters is going on for more than 100 years since the era of Taylor equation. The optimization of cutting parameters is confined to saving of cutting energy, which consists of less than 15% of total energy demand of machine tools. The remaining 85% of the energy is consumed as fixed energy by the auxiliary components (Sihag and Sangwan 2019) (coolant pump, axis feed motors, chip conveyor, spindle motor, router, fans, lights, lubricant pump, etc.). Therefore, analysis and reduction of fixed energy of the machine tools gained attention in the recent years (Jia et al. 2017; Mustafaraj et al. 2015; Li et al. 2011). The modeling and improvement strategies for cutting and fixed energy have been studied in recent years.

The studies have emphasized that efficient monitoring and transparency in energy flow of machine tools is a primary prerequisite for energy saving (Teiwes et al. 2018). It has been observed that energy assessment at each hierarchical level is essential to support efficient decision making for energy saving (Vikhorev et al. 2013). Therefore, innovative and effective energy monitoring and management approaches are required to promote energy efficient machining. Some attempts have been made towards energy data acquisition, mapping and analysis to save energy (Zein et al. 2011). A few studies have provided generic models for estimation of machine tool energy consumption using mathematical modeling and finite elemental analysis (Pervaiz et al. 2015; Huang et al. 2016). The correlation between the energy consumption and NC codes has been analyzed to obtain a better picture of energy consumption by various components of a machine tool (He et al. 2011). Micro analysis of the machining energy has been provided to quantify the energy consumption up to component and activity levels for better energy transparency (Sihag and Sangwan 2019). Event stream mapping has been used as an efficient approach for automated energy monitoring of the machine tools (Vijayaraghavan and Dornfeld 2010). But the application of intelligent techniques for energy disaggregation is hardly addressed. A few studies have used various machine learning approaches for determining the operating state of a machine tool (O'Driscoll et al. 2015; Sihag et al. 2018). The energy data is processed and classified into pre-defined classes using various classifiers such as support vector machine (SVM), tree classifier, k-nearest neighbor (k-nn), etc.

It is still a challenge for researchers to identify where and how the energy is consumed during a machining process due to complex structure of the machine tools and large number of energy consuming components. So far, the studies have been confined to the identification of the operational state of the machine tools. The existing research does not disaggregate the energy demand up to the component level to make it more transparent and easy to identify the focused measures for energy efficiency. To bridge this gap, the current study. The current study proposes a non-intrusive Electric-Load Intelligence (E-LI) system using a combination of load disaggregation algorithms (such as feature extraction, support vector machines, median absolute deviation etc.) to estimate the energy consumption and operational time for each machining state as well as for energy consuming components of a machine tool. The main contributions of the current study can be summarized as:

- Machine tool status identification using an Electric-Load-Intelligence system.
- Disaggregation of the energy consumption of machine tools up to the component level.
- Comparison of the accuracies of various classifiers used in the literature.

12.2 Methodology

12.2.1 Electric-Load Intelligence Concept

Electric-Load Intelligence (E-LI) is defined as a detailed analysis of energy consumption behavior of electric appliances and development of intelligent applications using structured algorithms to promote efficient use of electricity (Cheng et al. 2006). The proposed E-LI system serves as a novel solution for the widely discussed issue of industrial smart metering (Herrmann et al. 2011). It can be classified as a blind identification problem, where the energy consumption of individual components is identified using the aggregate data of the machine tool measured at the main power input. In general, the operating states of machine tools are classified as idle state, machine ready state and cutting state. The current study proposes a two-tier classification technique. In the first step, the energy and time consumption for each machining state is determined. In the next step, the energy and time consumed by each energy-consuming component in each machining state is calculated.

Mathematically, the power consumption of a machine tool can be represented as

$$P(t) = \sum_{i=1}^{n} P_t^i + \varepsilon_t \tag{12.1}$$

where $P(t)$ is the power consumed by machine tool at time t, P_t^i is the power drawn by component i at time t, n is number of energy consuming components and ε_t is the measurement error.

The main objective of the proposed algorithm is to decompose the total power consumption by the machine tool into component specific power signals in order to achieve disaggregated energy sensing. Each electrical component used in machine tools has a unique energy consumption pattern known as load signature. This signature is used to identify the specific component activation in the total power signature of the machine tool. The framework of the methodology used to develop the intelligent non-intrusive load monitoring (NILM) algorithm in the present study is presented in Fig. 12.1.

The power drawn by the machine tool is measured at the main supply input to obtain root mean square (RMS) power waveform. The activation of an event is identified using an event detection algorithm. A set of features is extracted for each event and compared against the feature data stored in the system library of trained classifiers. A series of conditional statements are applied to calculate the energy consumption and duration for each operational state and auxiliary components.

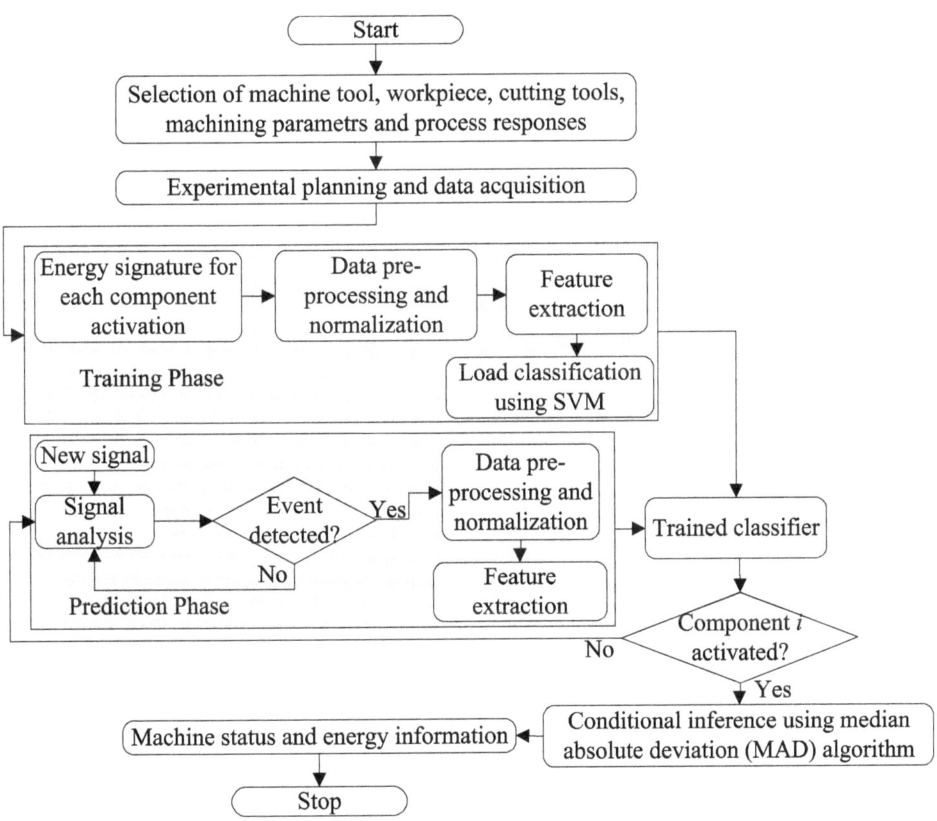

Fig. 12.1 The research methodology adopted for the present study

12.2.2 Load Signature Acquisition and Preprocessing

This study uses a supervised learning process to train the proposed E-LI system. The system is trained with a known data set labelled with a predefined class. For training the algorithm, the power signature for the major energy consuming components including spindle activation at different RPMs, activation of axis motors in x, y, and z directions at cutting and rapid speeds, rotation of automatic tool changer, activation of coolant pump, and basic module are recorded. The energy measurement device is installed at the main power input and load signature is recorded for each component activation for five times. The captured raw data is then processes with a series of data processing and conditioning techniques including filtering and normalization to reduce the noise, standardize the data and compensate for any power quality related issues.

12.2.3 Feature Extraction

The original recorded power signatures are raw waveforms with large dimensions and, generally not suitable for use by the classifiers. In practice, a finite set of measurements called as features are extracted from the raw waveform for each signal segment. The objective of feature extraction is to decrease the dimension of the raw signal while maintaining the required information. The bottom line to choose appropriate features is that they should be able to capture the similarities among the signals from the similar classes and differences among the signals from the different classes. In the present study, six unique features are extracted for each event activation. These features have been selected from literature on pattern recognition studies. They are later used by the different classification algorithms to assign class to a set of extracted features.

12.2.4 Load Disaggregation

In this step, activation of a component is classified using a known database of component power signatures and extracted features. Various classifiers have been used in the literature to classify the load signatures such as k-nn, SVM, Tree algorithm, etc. The selection of a suitable classifier depends on its performance and ability to classify the feature data into respective classes accurately. The present study compares the accuracy of these classifiers. SVM is selected as the suitable classifier for high accuracy as explained in Sect. 12.3. It is an efficient classification approach for high dimensional data and exhibits good performance in general (Widodo and Yang 2007). It uses a mapping function to project the non-linearly separable N-dimensional input vector x into a K-dimensional feature space (K > N) where data are linearly distinguishable.

12.2.5 *Conditional Inference for Detection of Material Removal*

Once the system is trained, it should be able to detect and classify any unknown/new load signature. As supervised learning approach covers only the predefined data classes, therefore, additional statements are required for identification of metal cutting process. Once the proposed E-LI system has identified and classified the spindle activation, machine tool is considered to be in the machine ready state. The system then searches for a rise in power consumption caused by the commencement of material removal. In the present study, median absolute deviation (MAD) algorithm is used to identify significant rise in power. At each point of time, the median is calculated and compared to previous value and cutting is identified when the change in median is higher than a predefined threshold. Once, the cutting operation is identified, the duration and energy consumption of machine ready and cutting state is calculated. Once trained, the algorithm can be run for any unknown/new data to predict the most likely class for the new data.

12.3 Experimental Validation of the Proposed E-LI System

12.3.1 *Experimental Set-Up*

The proposed E-LI system was verified on a 3-axis CNC vertical milling center (LMW KODI 40). A cuboidal aluminum block with dimensions 70 × 70 × 65 mm^3 was used as the test work piece. Multi-pass milling and drilling operations were performed for experimental validation of the proposed E-LI system. First, a layer of 3 mm thickness was removed from the surface using a 25 mm face mill. The cutting parameters selected for facing operation were: spindle speed = 2000 RPM, feed in x-direction = 50 mm/min and feed in z-direction = 100 mm/min. Next, a hole (Φ7 mm × 30 mm depth) was drilled at the work piece center using 7 mm drill. The cutting parameters selected for drilling operation were: spindle speed = 1500 RPM, feed in x-direction = 50 mm/min and feed in z-direction = 100 mm/min.

12.3.2 *Training Phase*

The energy meter was installed at the main energy supply of the machine tool. Fluke 435 series ii, three-phase power quality and energy analyzer was used to measure the energy. The key energy consuming components identified for the VMC are spindle activation at different RPM, activation of axis motors in x, y and z directions at cutting and rapid speeds, rotation of automatic tool changer, activation of coolant pump, and basic module. The power data for each component was recorded and stored for training of the algorithm. The power ratings for each component are also computed and recorded for energy calculations of individual components.

The recorded data is first filtered to reduce noise followed by normalization. A set of six features are obtained for the normalized signature of each activation and labelled. This data is used for training of the classifiers. In the present study, the classifiers used to classify the data are identified from the literature and their accuracy analysis is presented in Sect. 12.4.

12.3.3 Prediction Phase

Once the algorithm is trained and validated, it is able to detect and classify any new event activation. The proposed sensor starts to search for event activation using event detection algorithm. The system calculates the profile slope at each point and identifies an activation when slope rises beyond a predefined value k_1. The event is considered to be active until the slope of power profile decreases beyond a predefined threshold k_2. The power RMS values should also satisfy certain conditional statements to signify reduction in power consumption. The threshold slope values and power RMS conditions are determined based on preliminary experimentation and training. Once an event is identified, the power profile for that event is normalized. Six predefined features are calculated for the normalized profile and the event is classified using SVM classifier. The system uses the same approach to detect and classify each event during the milling process. After the coolant pump and spindle motor activation is identified and the system becomes stable, it looks for a rise in power consumption due to initiation of material removal process. Once the cutting starts, it keeps on recording the data until the cutting stops and power decreases beyond a predefined value. Similarly, the system records the start and stop of each component and machining state for the entire profile. A set of conditional statements are then used to calculate the energy consumption by each component and operational state. The power profile obtained for the multi pass face milling and drilling operations is shown in Fig. 12.2. The proposed algorithm detects two spindle activations, two coolant activations, and one tool change during the test run as shown in Fig. 12.2. It also identifies the machine ready and cutting states for the machine tool and quantifies the time and energy consumed by them. The total energy consumed for the machining process is hence divided into energy consumed in each machining state and further divided into energy consumed by each component in each state. Similarly, the time duration of machining process is disaggregated.

12.4 Accuracy Analysis of the Classifiers

Accuracy of a classification algorithm is an important criterion in selection of suitable classifier. In the present study, supervised learning data is used to obtain and compare the accuracies of different classifiers (Table 12.1). It is evident here that Quadratic

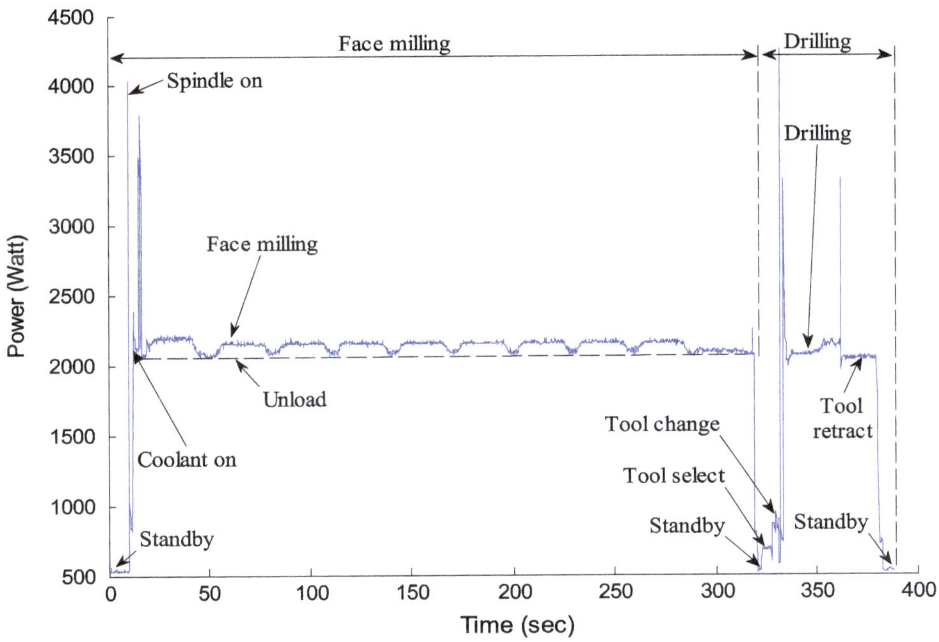

Fig. 12.2 Recorded power profile for the multi pass face milling and drilling operations

Table 12.1 Comparison of accuracy for various classifiers

Classifier	Accuracy (%)	Accuracy[a] (%)
Quadratic SVM	92.9	85.9
Cubic SVM	90.6	81.2
Linear SVM	91.8	82.4
Fine Gaussian SVM	87.1	82.4
Medium Gaussian SVM	90.6	83.5
Coarse Gaussian SVM	74.1	8.2
Linear discriminant	84.7	83.5
Quadratic discriminant	91.8	85.9
Fine k-nn	88.2	78.8
Medium k-nn	55.3	64.7
Weighted k-nn	87.1	81.2
Subspace k-nn	90.6	78.8
Subspace discriminant	83.5	83.5
Bagged trees	89.4	78.8
Medium tree	64.7	65.9

[a]Accuracy of the classifier using PCA explaining 95% variation

SVM exhibits the highest accuracy of 92.9%. The accuracy decreases to 85.9% with principal component analysis (PCA) at 95% explained variation.

The performance of the selected classifier is further analyzed using the confusion matrix as shown in Fig. 12.3. It provides an insight of the classifier performance by recording the number of times a sample is confused to be in another class. It helps to analyze the possible reasons for errors in classification results for a large number of labeled samples and makes the required adjustments in the classifier. Confusion matrix for the present study is shown in Fig. 12.3. It is observed that FX is often confused with FY and RX is confused with RY. The reason is that the axis motors for x and y directions have the same power ratings. Once the classifier is trained, a new set of data is provided to it for validation of its performance. Data set for eight events is provided to the classifier and it is observed that the events are classified accurately, for each case, by the trained classifier.

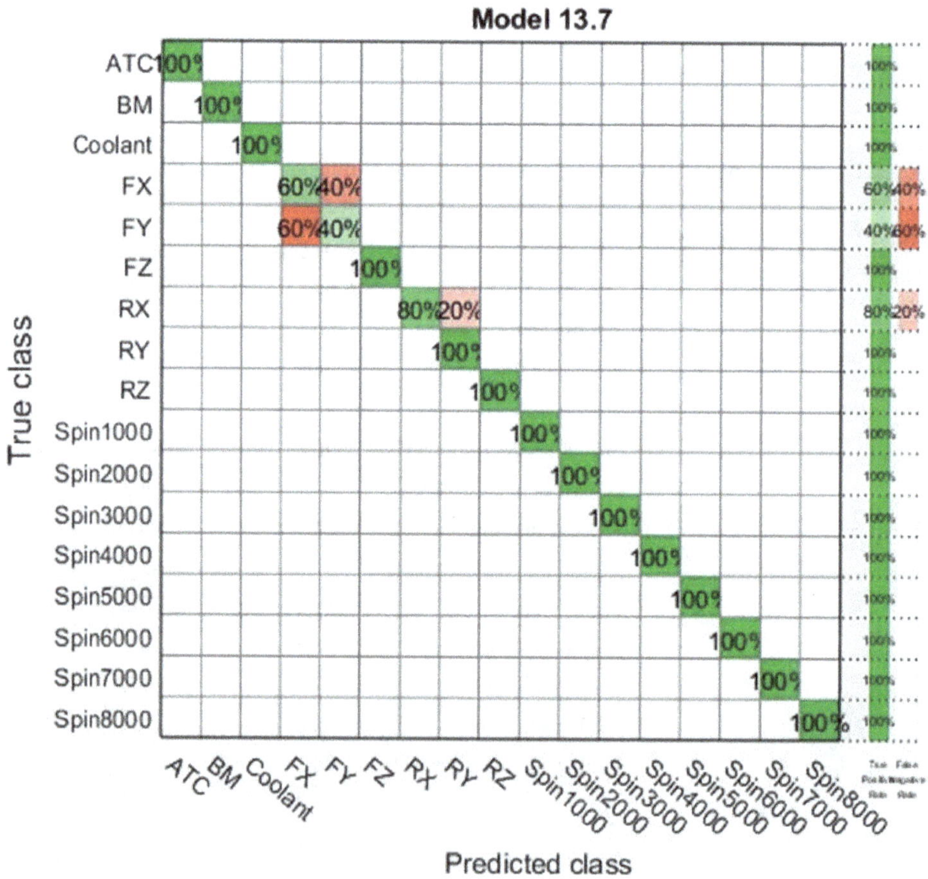

Fig. 12.3 Confusion matrix for SVM classifier

12.5 Results and Discussion

The proposed E-LI system calculates the total energy consumed and total time duration for the milling operation as 782.24 kJ and 368.50 s, respectively. The energy and time classification are shown in Figs. 12.4 and 12.5, respectively. The actual energy consumption calculated using the power profile is 782.11 kJ. It is evident from Fig. 12.4 that 72% of the total energy is consumed during the cutting state and out of this, only 7% is consumed for actual material removal. Major percentage of the total energy is consumed by the coolant pump (309.15 kJ) followed by basic module (205.91 kJ) and spindle motor (116.89 kJ). The reason behind this is that a high rated motor (1.1 kW) is used for coolant pump and it is active for 354.5 s. Basic module consists of the small components, which are essential for running the machine such as various relays, i/o data cards, PCBs, proximity/limit switches, lamps, etc. The power rating for basic module is less but high operation time results in higher energy consumption. It can be noted here that the present case study is conducted for a particular milling operation, hence, the idle time is negligible here. In actual industrial practices, the idle time of the machine tools is high. Hence, the energy consumed in the idle mode will be higher and the proportion of energy consumed for actual material removal will be even lesser. The disaggregated energy data provided by the proposed sensor can be used to detect the potential energy saving possibilities

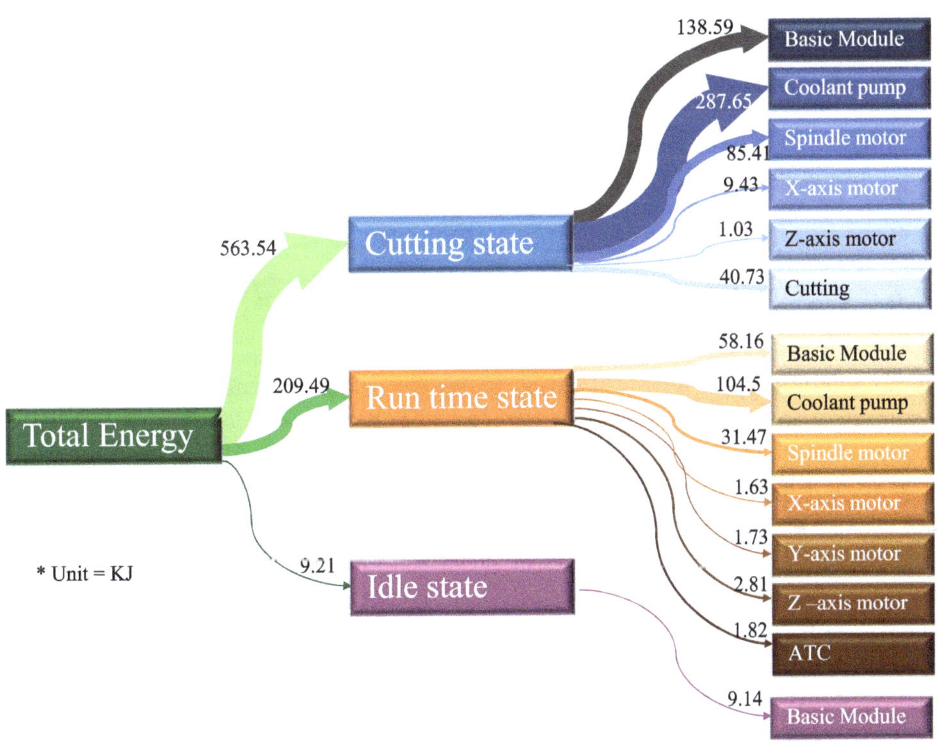

Fig. 12.4 Energy decomposition for multi pass face milling and drilling experiments

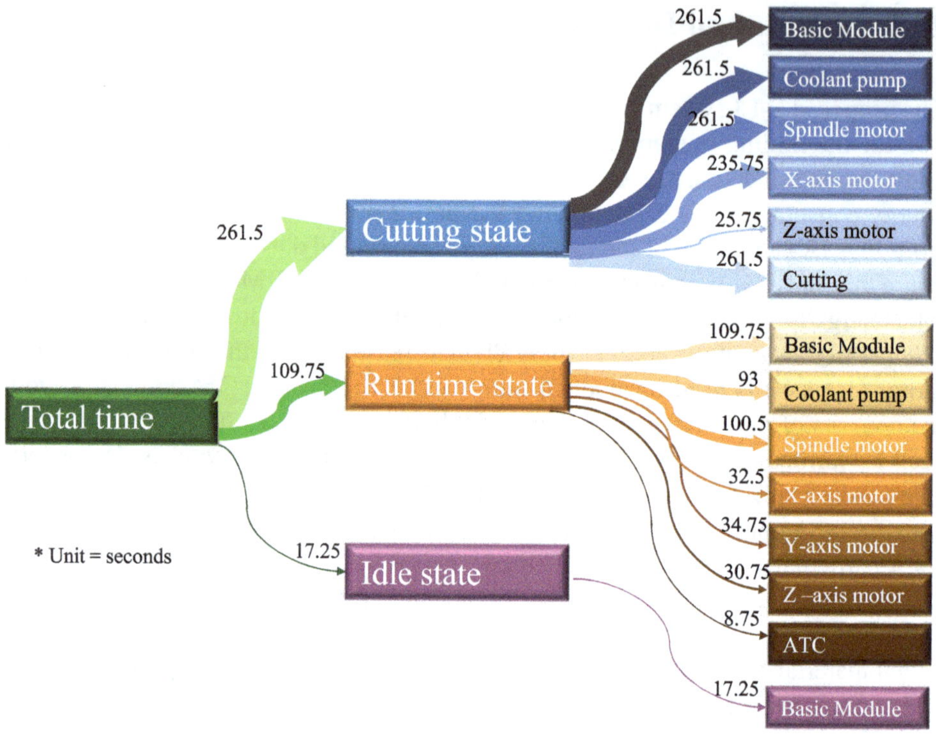

Fig. 12.5 Processing time decomposition for multi pass face milling and drilling experiments

and promote use of more efficient components for improved energy performance of the machine tools. The disaggregation up to component level provides better energy flow transparency leading to better decisions for energy efficiency.

12.6 Conclusions

Disaggregation of machine tool energy profile up to component level is an essential requirement to achieve better transparency in energy flow. However, it is a challenging issue for researchers due to the complexity of the machine tools. This study presents an Electric-Load Intelligence (E-LI) system for disaggregation of machine tool energy profile up to operating state and component level using a supervised learning approach. It is a non-intrusive energy monitoring method, which computes the activation time and energy consumed by each energy-consuming component by analyzing the power profile measured at the main power inlet of the machine tool. It is a low cost and viable alternative to expensive and complex multi-sensor data acquisition systems. The accuracies of different classifiers commonly used are compared and it is found that quadratic SVM classifier has maximum accuracy. The proposed methodology is verified through a milling case study. The results reveal that the total energy consumed for the test run is 782.24 kJ, but the energy consumed for actual material removal is only 40.73 kJ. The coolant pump consumes 392.15 kJ energy, which is almost half of the total energy consumption. The coolant pump is identified as an inefficient component here and should be replaced with a more efficient pump. It is concluded that the proposed E-LI system is an efficient and feasible approach to identify the time and energy requirement at each operating state and individual components. It provides better transparency in the energy consumption behavior of machine tools to motivate the behavioral changes leading to the reduction in energetic impacts of machining.

References

Camposeco-Negrete C (2015) Optimization of cutting parameters using response surface method for minimizing energy consumption and maximizing cutting quality in turning of AISI 6061 T6 aluminum. J Clean Prod 91:109–117

Cheng JWM, Kendall G, Leung JSK (2006) Electric-load intelligence (E-LI): concept and applications. In: ENCON 2006–2006 IEEE region 10 conference. IEEE, pp 1–4

He Y, Liu F, Wu T, Zhong FP, Peng B (2011) Analysis and estimation of energy consumption for numerical control machining. Proc Inst Mech Eng Part B J Eng Manuf 226:255–266

Herrmann C, Suh SH, Bogdanski G et al (2011) Context-aware analysis approach to enhance industrial smart metering. In: Hesselbach J, Herrmann C (eds) Glocalized solutions for sustainability in manufacturing. Springer, Berlin, Heidelberg, pp 323–328

Huang J, Liu F, Xie J (2016) A method for determining the energy consumption of machine tools in the spindle start-up process before machining. Proc Inst Mech Eng Part B J Eng Manuf 230:1639–1649

Jia S, Tang R, Lv J, Yuan Q, Peng T (2017) Energy consumption modeling of machining transient states based on finite state machine. Int J Adv Manuf Technol 88:2305–2320

Kant G, Sangwan KS (2014) Prediction and optimization of machining parameters for minimizing power consumption and surface roughness in machining. J Clean Prod 83:151–164

Li W, Kara S (2011) An empirical model for predicting energy consumption of manufacturing processes: a case of turning process. Proc Inst Mech Eng Part B J Eng Manuf 225:1636–1646

Li W, Zein A, Kara S, Herrmann C (2011) An investigation into fixed energy consumption of machine tools. In: Hesselbach J, Herrmann C (eds) Glocalized solutions for sustainability in manufacturing. Springer, Berlin, Heidelberg, pp 323–328

Mustafaraj G, Cosgrove J, Rivas-Duarte MJ, Hardiman F, Harrington J (2015) A methodology for determining auxiliary and value-added electricity in manufacturing machines. Int J Prod Res 53:5265–5277

O'Driscoll E, Kelly K, O'Donnell GE (2015) Intelligent energy based status identification as a platform for improvement of machine tool efficiency and effectiveness. J Clean Prod 105:184–195

Pervaiz S, Deiab I, Rashid A, Nicolescu M (2015) Prediction of energy consumption and environmental implications for turning operation using finite element analysis. Proc Inst Mech Eng Part B J Eng Manuf 229:1925–1932

Sihag N, Sangwan KS (2019) An improved micro analysis-based energy consumption and carbon emissions modeling approach for a milling center. Int J Adv Manuf Technol 1–17

Sihag N, Sangwan KS, Pundir S (2018) Development of a structured algorithm to identify the status of a machine tool to improve energy and time efficiencies. Procedia CIRP 69:294–299

Teiwes H, Blume S, Herrmann C, Rössinger M, Thiede S (2018) Energy load profile analysis on machine level. Procedia CIRP 69:271–276

Vijayaraghavan A, Dornfeld D (2010) Automated energy monitoring of machine tools. CIRP Ann Manuf Technol 59:21–24

Vikhorev K, Greenough R, Brown N (2013) An advanced energy management framework to promote energy awareness. J Clean Prod 43:103–112

Widodo A, Yang BS (2007) Support vector machine in machine condition monitoring and fault diagnosis. Mech Syst Signal Process 21:2560–2574

Zein A, Li W, Herrmann C, Kara S (2011) Energy efficiency measures for the design and operation of machine tools: an axiomatic approach. In: Proceedings of the 18th CIRP international conference on life cycle engineering, Springer, Berlin, Heidelberg, pp 274–279

Zhang L, Zhang B, Bao H, Huang H (2018) Optimization of cutting parameters for minimizing environmental impact: considering energy efficiency, noise emission and economic dimension. Int J Precis Eng Manuf 19:613–624

13

Magnesium Metal Matrix Composites and their Machining: A Study of Energy Analysis

Navneet Khanna and Prassan Shah

Abstract Adoption of sustainable machining techniques shall offer the local industry a cost-effective route to improve its environmental, economic and social footprint when it comes to machine difficult-to-cut materials. This experimental study investigates the behavior of sustainable cutting fluid approaches on active cutting energy (ACE), active energy consumed by machine tool (AECM) and energy efficiency (EE) for machining PMMCs (particulate metal matrix composites) of magnesium at different combinations of rotational speed and feed. Minimum Quantity Lubrication (MQL), cryogenic and CryoMQL machining are performed on in-house developed MQL and cryogenic experimental setups and the results obtained from them are compared with dry machining. The L_{36} orthogonal array is employed to design the experiments. It is observed that cryogenic machining consumes comparatively lower ACE and AECM among the four cutting fluid approaches. It is found that dry machining provides comparatively lower EE among four cutting fluid approaches. From the main effects plot, it is observed that cryogenic assistance further improves the machining performance of the MQL technique and offers better EE. The results of Analysis of Variance (ANOVA) suggest that rotational speed, cutting fluid approach and feed are the significant parameters that affect the EE in descending order respectively.

Keywords Cryogenic machining · Energy efficiency · Particulate metal matrix composites

---

N. Khanna (✉) · P. Shah
Advanced Manufacturing Laboratory, Institute of Infrastructure Technology Research and Management, Ahmedabad 380026, GJ, India
e-mail: navneetkhanna@iitram.ac.in

P. Shah
e-mail: prassan.shah.17pm@iitram.ac.in

13.1 Introduction

It has been forecasted by International Energy Agency that up to 2030 the demand for electrical energy increases by 1.7% per year. Manufacturing processes consume about 30% of the total electrical energy produced. Almost all machining processes being a subpart of manufacturing process consume electrical energy for performing work. By considering this fact, it is required to consider energy consumption as an important machinability indicator along with tool wear, cutting force and surface roughness for machining processes (Bilga et al. 2016; Li and Kara 2011). Due to lightweight and exceptional mechanical properties at elevated temperature, the PMMCs of Mg are widely used in the aerospace and automotive industries (Khanna et al. 2019). Machinability of PMMC is poor due to the existence of hard particles within a softer metal matrix. To combat it, cutting fluid is used which reduces the cutting zone temperature, cutting force, surface roughness and tool wear. MQL and cryogenic machining have a low impact on the environment due to a minimum and no usage of cutting oil respectively during machining. These processes do not only decrease the hazards to the operator but also eliminate the chip recyclability process (Khanna and Agrawal 2020; Adler et al. 2006). Though the dry machining does not consume any type of cutting fluid, it is not sustainable because it generates higher surface roughness and tool wear and hence results in lesser product quality and productivity (Canter 2009). So, it will be interesting to compare the above-mentioned sustainable processes with dry machining. In this context, the EE is measured for dry, MQL, cryogenic and CryoMQL machining processes at different combinations of rotational speed and feed. EE is defined as the ratio of ACE to AECM (Bilga et al. 2016). ACE is the net energy consumed during the cutting process while the AECM includes the ACE and the energy losses occurred due to mechanical transmission, electrical motors and electrical networks. The difference between active power consumed by machine tool (APCM) with material removal and without material removal i.e., when the spindle is on but no contact between cutting tool and the workpiece is active cutting power (ACP) and it converts into ACE when it is multiplied by cutting time (Bilga et al. 2016).

Pu et al. (2012) compared surface integrity of machined surface of the AZ31B Mg alloy for dry and cryogenic machining. A combination of large tool radius and cryogenic machining provided higher values of compressive stresses as compared to dry machining.

Kara and Li (2011) made an empirical model of specific energy consumption (SCE) for turning and milling operations for dry and wet machining. The lower values of SCE were observed in the dry machining as compared to wet machining for the same material removal rate.

Madanchi et al. (2019) developed a model to identify the effect of cutting fluid strategies with the change in process parameters (cutting speed, feed and depth of cut) on the energy consumption and cost. This model considered the correlation of elements of the machining system with the change in the cutting fluid strategies.

From the above literature, it can be inferred that the selection of cutting process parameters and cutting fluid influence the machining performance and eventually energy consumption and EE during the machining process.

The industry is facing a decisive challenge to make energy-efficient machine tools for machining difficult-to-machine materials. Research work focused on this theme is required to increase pertinent understanding in order to develop an energy-efficient hybrid machining facility for the local industry. The adoption of sustainable production techniques shall allow the local industry a cost-effective way to fulfill its socio-economic and environmental challenges. In relation to this, the present work compares the effect of process parameters on EE for dry, MQL, cryogenic and CryoMQL machining. It is envisioned that the findings of this work will help in the development of optimized in-house retro-fitted hybrid machining facility.

13.2 Experimental Setup and Design of Experiments

AZ91/5SiC PMMC is used in the form of a 20 mm diameter and 190 mm length rod for turning tests on a conventional lathe. In the final composition of AZ91/5SiC PMMC, 5% SiC is reinforced with 67 μm particle size in the metal matrix of AZ91.

For MQL machining, in-house developed mist generator is used. LRT30 cutting oil is used as a lubricant with 14 ml/h flow rate. For performing cryogenic machining, LN_2 was stored in Dewar at 6 bar pressure. To convey the N_2 in liquid form from Dewar to the cutting zone, a vacuum insulated hose pipe is used. In MQL and cryogenic machining, 2 mm diameter nozzle is used. For CryoMQL machining, the above two setups are merged in such a way that MQL and LN_2 stroke on flank and rake face of the cutting tool respectively. Figure 13.1 describes the experimental setup of CryoMQL machining.

Here CNMG120404AH DLC (Diamond Like Coating) insert is used with MCLNR2020K12 tool holder. For every experiment, a fresh cutting edge is used to have the same experimental treatment. Fluke 435 (series-II) 3 phase energy and power quality analyzer is used to measure the APCM. To limit the experimental design, the rotational speed, feed and cutting fluid approach is considered as factors

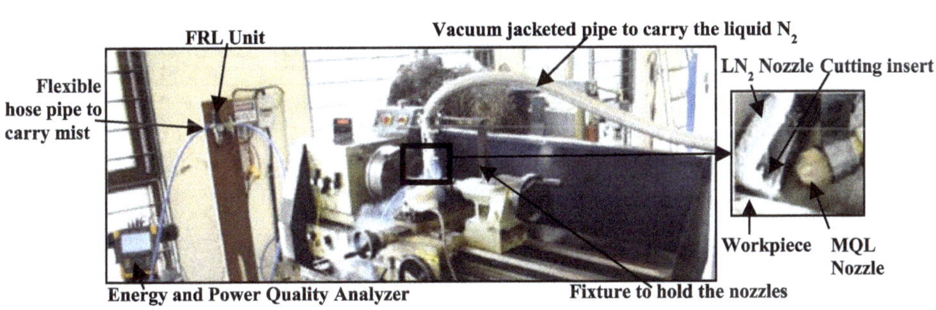

Fig. 13.1 Experimental setup for CryoMQL machining

(Bilga et al. 2016; Kara and Li 2011). Here L_{36} orthogonal array is used which consists of three levels of factors namely rotational speed (835 (N1), 557 (N2), 371 (N3) rpm), feed (0.111 (F1), 0.222 (F2), 0.333 (F3) mm/rev) and four levels of cutting fluid approach (dry, MQL, cryogenic and CryoMQL) to analyze the results of ACE, AECM, and EE.

13.3 Results and Discussion

The results of ACE, AECM, and EE are shown in Fig. 13.2. It is observed from Fig. 13.2 that dry machining gives higher values of ACE, AECM and lower values of EE at most of the tests among four cutting fluid approaches. For cryogenic machining, ACE and AECM are found to be the lowest at most of the tests among four cutting fluid approaches except at lower rotational speed and higher feed. The reason for lower energy consumption for cryogenic machining may be a noteworthy grain refinement of Mg alloy on the surface of machined parts at low temperature. It reduces the requirement of cutting force and hence energy (Pu et al. 2012).

It has been observed that the main effects and interaction plots provide qualitative information regarding the impact of factors on response with direction. Figures 13.3 and 13.4 show the main effects and interaction plots for EE respectively. From Fig. 13.3, it is observed that rotational speed, cutting fluid approach and feed are the significant parameters that affect the EE in descending order respectively. It is evident that as the rotational speed increases the EE increases rapidly. Though there is a marginal difference of EE for the MQL, cryogenic and CryoMQL machining, they have significantly higher value of EE as compared to dry

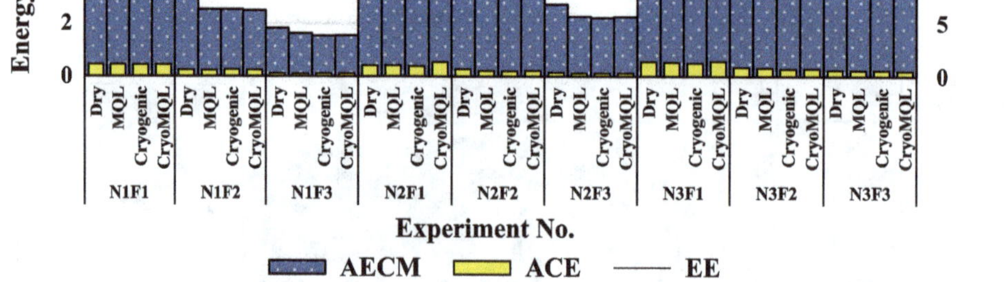

Fig. 13.2 Results of AECM, ACE, and EE for different cutting fluid approaches at different cutting parameters

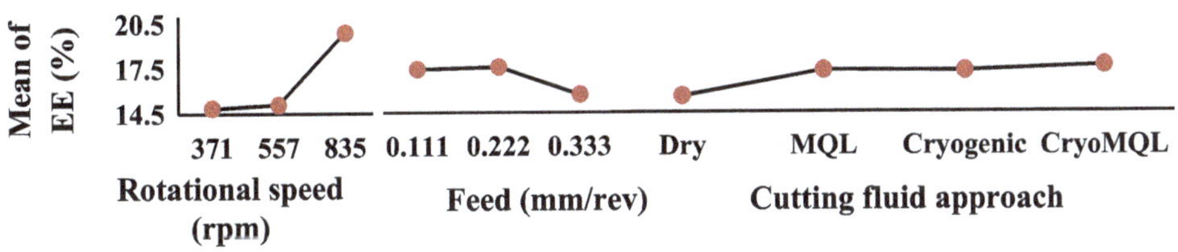

Fig. 13.3 Main effects plot for EE

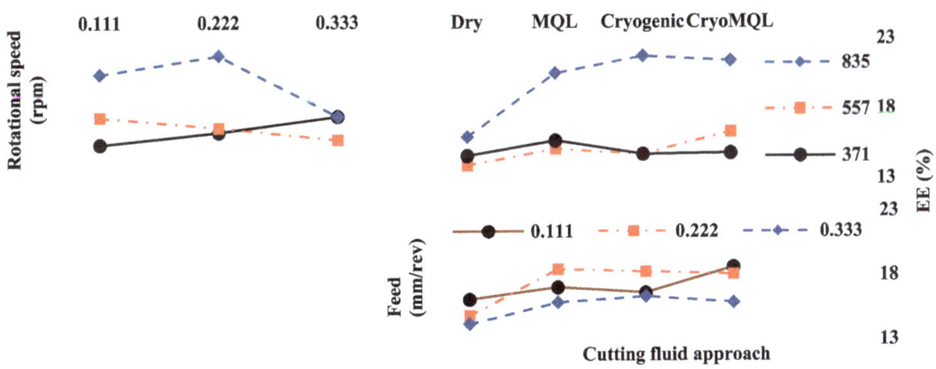

Fig. 13.4 Interaction plot for EE

machining. It is also observed that at intermediate feed (0.222 mm/rev) the value of EE is higher. This result does not provide sufficient information if the interaction effect of factors is not considered.

From the interaction plot (Fig. 13.4), it is clear that rotational speed with feed and cutting fluid approach with rotational speed strongly affect the EE if the interaction effect of factors is considered. It is also clear from the Fig. 13.4 that for all cutting fluid approaches if the value of rotational speed increases the value of EE increase rapidly.

ANOVA is considered as a valuable tool to predict the impact of factors on response quantitatively with interaction effect (Khanna and Agrawal 2020). Here the p-test is performed with ANOVA to analyze the information regarding the significance of parameters on response. For any parameter having p-value less than 0.05 is considered as a significant parameter with 95% confidence. From the Table 13.1, it is reconfirmed that rotational speed, cutting fluid approach and feed are the significant parameters, which affect the EE by 43.21%, 12.23%, and 5.01% respectively.

In this study, only the spindle network is considered as machining system and the rest of the elements are not included (e.g. air-compressor used for the MQL and CryoMQL machining). If the energy consumed by air-compressor is considered then even lower values of ACE and AECM must be observed in cryogenic machining as compared to MQL and CryoMQL machining. This clearly establishes cryogenic machining as eco-efficient machining.

Table 13.1 Results of ANOVA for EE

Source	DF	Seq SS	Adj SS	Adj MS	p	Contribution (%)
Rotational speed (rpm)	2	174.04	177.16	88.58	0.000	43.21
Feed (mm/rev)	2	20.17	21.67	10.83	0.040	5.01
Cutting fluid approach	3	49.25	31.83	10.61	0.030	12.23
Rotational speed (rpm) * Feed (mm/rev)	4	70.41	75.32	18.83	0.000	17.49
Feed (mm/rev) * Cutting fluid approach	6	9.60	10.17	1.69	0.710	2.38
Rotational speed (rpm) * Cutting fluid approach	6	46.05	46.05	7.67	0.060	11.44
Error	12	33.15	33.15	2.76		8.24
Total	35	402.71				

13.4 Conclusions

This study presents cryogenic machining as an eco-efficient machining technique suitable to reduce energy consumption in modern-day manufacturing processes. Trials are carried out using in-house developed retro-fitted hybrid machining facilities. The following conclusions are drawn from the study.

- The lowest values of ACE and AECM at most of the turning tests have been obtained for cryogenic machining among four cutting fluid approaches. Cryogenic assistance further improves the machining performance of the MQL technique and offers higher values of the EE.
- From the results of the main effects plot, rotational speed, cutting fluid approach and feed are the significant parameters affecting the EE in descending order respectively. With the interaction effect of parameters, it is clear that for all four cutting fluid approaches higher values of EE are found at higher rotational speed (835 rpm). It is also clear that the interaction effect of rotational speed with feed and cutting fluid approach with rotational speed affect the EE strongly.
- From the results of ANOVA, it is observed that rotational speed, cutting fluid approach and feed affect the EE by 43.21%, 12.23%, and 5.01% respectively.

Acknowledgements The authors would like to thank the SERB-DST, Government of India, for the financial support given under the Project (ECR/2016/000735) titled "Design and Development of Energy Efficient Cryogenic Machining Facility for Heat Resistant Alloys and Carbon Fibre Composites". Authors are also thankful to Prof. N. M. Suri (Punjab Engineering College, Chandigarh) for sharing the in-house developed AZ91/5SiC PMMC material to evaluate its machining performance.

References

Adler DP, Hii WW, Michalek DJ, Sutherland JW (2006) Examining the role of cutting fluids in machining and efforts to address associated environmental/health concerns. Mach. Sci. Technol 10(1):23–58. https://doi.org/10.1080/10910340500534282

Bilga PS, Singh S, Kumar R (2016) Optimization of energy consumption response parameters for turning operation using Taguchi method. J Clean Prod 137:1406–1417. https://doi.org/10.1016/j.jclepro.2016.07.220

Canter N (2009) The possibilities and limitations of dry machining. Tribol Lubr Technol 65:40–44

Kara S, Li W (2011) Unit process energy consumption models for material removal processes. CIRP Ann Manuf Technol 60:37–40. https://doi.org/10.1016/j.cirp.2011.03.018

Khanna N, Agrawal C (2020) Titanium machining using indigenously developed sustainable cryogenic machining facility. In: Gupta K (eds) Materials forming, machining and post processing. Materials forming, machining and tribology. Springer, Cham

Khanna N, Suri NM, Agrawal C, Shah P, Krolczyk GM (2019) Effect of hybrid machining techniques on machining performance of in-house developed Mg-PMMC. Trans Indian Inst Met 72(7):1799–1807. https://doi.org/10.1007/s12666-019-01652-w

Li W, Kara S (2011) An empirical model for predicting energy consumption of manufacturing processes: a case of turning process. Proc Inst Mech Eng Part B J Eng Manuf 225(9):1636–1646. https://doi.org/10.1177/2041297511398541

Madanchi N, Thiede S, Gutowski T, Herrmann C (2019) Modelling the impact cutting fluid strategies on environmentally conscious machining systems. Procedia CIRP 80:150–155. https://doi.org/10.1016/j.procir.2019.01.068

Pu Z, Outeiro JC, Batista AC, Dillon OW Jr, Puleo DA, Jawahir IS (2012) Enhanced surface integrity of AZ31B Mg alloy by cryogenic machining towards improved functional performance of machined components. Int J Mach Tools Manuf 56:17–27. https://doi.org/10.1016/j.ijmachtools.2011.12.006

14

Eco-Industrial Networks: Analysis and Design

Manbir Sodhi, Fabian Schulze, Callum Bissett, and Mercedes Rivero-Hudec

Abstract The development of industrial clusters has, until recently, been guided solely by economic considerations. Recent advances in measuring and assessing the environmental and social consequences of industrial production have revealed the impact of human activity on the environment and society. Eco-Industrial Networks (EINs) are being viewed as a possible solution for reducing the environmental consequences of industrial production. In this paper, we review existing approaches used for analyzing and designing EINs. Based on salient works, design requirements for models to facilitate the organization of EINs are presented and discussed.

Keywords Eco-Industrial Networks · Industrial Symbiosis · Industrial Ecology

14.1 Introduction

Increased attention is being paid to ecosystems and the consumption of resources. As a result, governments and commercial organizations are supporting shifts towards sustainable development. There are many examples of this: companies are being held liable for their environmental impact in many regions all over the world (Abdullah et al. 2019; Roberts 2004) and some countries, for instance China, have gone a step further by promoting a comprehensive legal strategy called circular economy (Yuan et al. 2006). When proposing the concept of industrial ecology, Frosch and Gallopoulos (1989) suggested a holistic approach for companies to efficiently achieve improvements in all three dimensions of sustainable development, economy, environment, and society, suggesting that industrial systems should operate like natural eco-systems (Frosch 1994; Allenby 1992; Jelinski et al. 1992). The waste and byproducts of one company could be the inputs of another (Abdullah et al. 2019). This can be done by setting up Eco-Industrial Parks (EIPs) or Networks (EINs), which involve the cooperation of companies and communities sharing and using their resources and byproducts, synergistically reducing waste. Economies of scale

M. Sodhi (✉) · F. Schulze · C. Bissett · M. Rivero-Hudec
Department of Industrial Engineering, University of Rhode Island, Kingston, RI 02881, USA
e-mail: sodhi@uri.edu

can help to achieve economic improvements (Tudor et al. 2007). The case of Kalundborg in Denmark is one of many promising examples (Ehrenfeld 1997; Bain et al. 2010). However, Drexhage and Murphy (2012) and Gu et al. (2013) claim that this concept exists mainly in theory and that successful cases are not set up from scratch, however have developed in response to fortunate circumstances. Although there is governmental and private interest for designing and operating industrial parks or networks, the literature does not reveal much in the way of mathematical and computational modeling for the improvement and design of corporative networks. The development of such models can help to support the process of turning the key concepts of the idea of industrial ecology by providing optimal decisions, assessing patterns, and investigating key factors in the design and operation of EINs (Gu et al. 2013). This paper aims to map the mathematical approaches used for EIN design and indicate how they may be used for designs in the future.

14.2 Literature Review and Modeling Requirements

Eco-Industrial Networks and related concepts have been identified by many different authors over the last two decades (Gu et al. 2013; Veiga and Magrini 2009; Lowe 2001; Schlarb 2001; Chertow 2000; Côté and Cohen-Rosenthal 1998; Côté and Hall 1995). The principal related definitions are:

Eco-industrial park (EIP): "An eco-industrial park or estate is a community of manufacturing and service businesses located together on a common property" (Lowe 2001). Eco-industrial Networks: EIPs without geographic proximity (Ehrenfeld and Chertow 2002). The terms "industrial symbiosis networks" (Domenech and Davies 2011), and "zero waste networks" (Curran and Williams 2012) are interchangeable. The participants of a geographically spread virtual EIP are EINs. Roberts defined EINs as "networks of EIPs at national or global levels" (Roberts 2004).

An appraisal of the literature on the application of modeling methods and optimization approaches used in the field of Industrial Ecology as related to EINs has been conducted. The search was conducted with Google Scholar primarily, using terms such as: "EIN", eco industrial parks", "circular economy" with "mathematical model", "optimization" etc. From the resultant list, salient articles were identified based on the type of solution methodology used, and these have been reported in this paper. These approaches are clustered (Table 14.1) based on the method used and classified according the discussion in the previous paragraphs. Some proposed models may be part of more than one cluster. The cluster bi level fuzzy optimization requires, for example, a fuzzy optimization, and mixed-integer linear programming (MILP) or mixed-integer non-linear programming (MINLP), which are clusters themselves. The clusters represent main practices of approaching modeling of industrial ecology.

Walther et al. (2008), and Miettinen (1999) develop a negotiation algorithm for the coordination of material flow in recycling networks. The idea of industrial ecology has not been mentioned in these publications; however, based on mathematical models and an interactive negotiation algorithm, new symbiosis can be created. In

Table 14.1 Modeling clusters for eco-industrial parks

Cluster	Main publication
Input-Output analysis	Ayres and Ayres (2002), Duchin (1992)
Material flow analysis	Bringezu et al. (1997), Bringezu and Moriguchi (2002), Bringezu and Moriguchi (2018), Lee et al. (2006)
Mixed-integer linear programming	Boix et al. (2015), Gonela and Zhang (2014), Karlsson and Wolf (2007), Geng et al. (2016)
Lagrange relaxation and penalty functions	Walther et al. (2008), Pishvaee et al. (2009)
Multi-objective optimization	Gu et al. (2013), Azapagic and Clift (1999), Erol and Thöming (2005)
Fuzzy optimization	Taskhiri et al. (2011)
Bi-level optimization	Geng et al. (2016), Kastner et al. (2015), Chew et al. (2009)
Evolutionary optimization	Theo et al. (2016)
System dynamics and complex network theory	Kuznetsova et al. (2017), Mantese and Amaral (2017)
Agent-based modeling	Romero and Ruiz (2014), Romero and Ruiz (2013)

order to solve the optimization model, Lagrange relaxation is applied. While the original objective function contains economic measurements, the Lagrange relaxation allows a variation of the recycling rate and thus accounts for environmental issues. Penalty functions are a common multi-objective optimization method (Miettinen 1999). Pishvaee et al. (2009) provide a meta-investigation of modeling approaches for reverse and integrated networks considering uncertainty.

While the subject matter of this investigation is a sustainable reverse logistics network, environmental targets or the idea of industrial ecology have not been applied. Different methodologies have been developed in the literature in order to solve multi-objective optimization models (Walther et al. 2008). A successful application of the NIMBUS (non-differentiable interactive multi-objective bundle-based optimization system) method to optimization of eco-industrial parks has been proposed by Gu et. al. (2013); they apply the whole process of an interactive multi-objective optimization to both eco-industrial park design and optimization; this composition of computational and mathematical modeling considers multiple waste product flows. Since this tool is web-based, it is considered to have a high usability. Erol and Thöming (2005) consider multiple stakeholders with an interactive negotiation framework, neglecting social performance; uncertain behavior is also not considered. This methodology supports the improvement and design of eco-industrial parks. Li et al. (2009) consider chemical processes in general for industrial ecology; they apply TOPSIS (technique for order preference by similarity to ideal solution) and solve with an NSGA-II (non-dominated sorting generic algorithm). Erol and Thöming (2005) combine the simultaneous analysis of environmental impact sensitivity (SAEIS) with multi-objective optimization performed by MINLP; they model the trade-off between economy and environment under consideration of life cycle analysis (LCA) factors. Li et al. (2009) provide the basis for this approach by illustrating the application of LCA to process optimization. The interactive surrogate worth trade-off method (ISWT) has only been applied to power plants, not to EIPs.

Taskhiri et al. (2011) suggest a model to achieve a compromise among the potentially conflicting fuzzy goals of the various EIP stakeholders. Unlike the following approach, this mathematical optimization model does not consider a hierarchical structure. Karlsson and Wolf (2007) consider the hierarchy of decision-making in an eco-industrial park using a bi-level fuzzy optimization; they take into account the participating plants by means of an individual fuzzy cost goal while the upper level and overall goal of an EIP authority is the minimization of resource consumption and generation of waste. Their model thus includes environmental and economic targets but does not consider social issues.

Using a fuzzy function, lower and upper boundaries are included and provide a range in which alternative economic outcomes are acceptable for participants. The bi-level especially considers the hierarchy of stakeholders. It applies the Stackelberg Game to mathematical optimization. The basic idea of applying game theory approaches has also been investigated by Chew et al. (2009). This multi-objective bi-level optimization has also been applied to other problems such as transport planning and management problems in the past, and Kastner et al. (2015), and Yin (2002) use a nonlinear solver to find an optimal solution of an example case. The application of the max-min-concept seeks to maximize the satisfaction of the least satisfied company.

Huo and Chai (2008) set up a simulation to understand evolution of industrial ecology patterns and provide new implications for design, improvement, and prediction of structural evolutions. They investigate patterns and apply evolutionary principles as well as nonlinear partial differential equations with boundary conditions and thus computationally implement interacting organisms. Evolutionary algorithms solve many nonlinear programs, and are can be used to solve multi-objective optimization problems (Zitzler and Thiele 1999). However, other than Qu et al. (2014), most of the nonlinear programs have an underlying mathematical model to be solved.

Romero and Ruiz (2013, 2014) propose the application of agent-based modeling to the optimization and design of eco-industrial parks. Single companies are implemented as agents with an individual behavior and an individual economic and ecological goal (Bichraoui et al. 2013). Among other notable approaches, LCA is product based and not based on company level approach for handling eco-industrial models. However, Tong et al. (2013) applies the LCA to a system for water reuse in an industrial park. In order to determine the correct partners for increasing competitive advantage, many mathematical programming models, such as linear programming (Pan 1998), stochastic integer programming (Afshari et al. 2016), and multi-objective programming (Huang et al. 2010) have been proposed.

14.3 Analysis and Requirements

Based on the literature review, eight requirements for modeling EINs can be identified. Models should be able to include (i) Economic objectives; (ii) Environmental objectives; (iii) Social objectives; (iv) Multiple flows; (v) Multiple stakeholders; (vi) Negotiation of alternatives; (vii) Uncertainty; (viii) Ability to generate optimal solutions; and (ix) Usability. The relative importance of the requirements is dependent on the particular situation, but a comprehensive model for an EIN should address all these requirements.

The development potential of the different approaches is shown in Table 14.2 by four categories: A, B, C and D. Category A: for analysis usefulness; B: for improvements; C: for extending capabilities; and D: for design of Eco-Industrial Networks. From the table, if the goal of the model being developed relates to metrics associated with multiple flows, Input-Output analysis, Material Flow and Lagrangian Penalty based methods can be used for the purpose. Similarly, based on the development potential, Input-Output analysis, Material Flow can be used for analyzing exchanges in Eco-Industrial Networks. Likewise, Mixed Integer Programming, Agent based approaches can be used for designing new EINs. Multicriteria methods, Fuzzy approaches and Bilevel programming have been used for improving existing networks, and, Evolutionary and System Dynamics methods can be used to extend/augment existing EINs either by extending the network size or by including additional materials.

14.4 Summary

Some special patterns have been discovered in this review of mathematical approaches for modeling and designing EINs. There is no current method for achieving an optimized decision for creating new eco-industrial parks and networks. The only approach providing such an idea has been proposed by Romero and Ruiz (2013, 2014). While nearly every modeling approach that has been applied by a publication in the field of industrial ecology directly considers economic and ecological performance indicators in the objective functions, social performance has not been addressed. It is noteworthy that most of the publications only consider a single flow of material in a network; useful by-products are rarely considered explicitly by any model.

Based on the review, several design requirements for modeling EINs are determined. The domain requirements and the modeling techniques best suited for each domain are mapped. Future work on the use of these requirements for designing EINs for developing economies is ongoing.

Table 14.2 Mapping techniques to modeling techniques

No.	Objective/domain	Input-Output	Material flow	Mixed integer LPS	Lagrangian penalty	Multicriteria	Fuzzy	Bilevel/fuzzy	Evolutionary	System dynamics	Agent-based	Desired approach
1	Economic	×	✔	✔	✔	✔	✔	✔	✔	✔	✔	✔
2	Environmental	×	✔	✔	✔	✔	×	✔	✔	✔	✔	✔
3	Social	×	×	×	×	×	×	×	×	×	×	✔
4	Multiple flows	✔	✔	×	✔	×	×	×	✔	×	×	✔
5	Multi-stakeholders	×	×	×	×	✔	✔	✔	×	✔	✔	✔
6	Negotiation/alternatives	×	×	✔	✔	✔	✔	✔	✔	×	×	✔
7	Uncertainty	×	×	✔	✔	×	×	×	✔	×	✔	✔
8	Optimality and uniqueness	×	×	✔	✔	×	✔	✔	×	×	×	✔
9	Usability	×	✔	✔	✔	×	×	✔	×	×	×	✔
Development potential		A	A	D	C	B	B	B	C	C	D	D

References

Abdullah R, Malik E, Pratiwi ET, Abdullah L, Sulili A., Heliawaty (2019) Influence of corporate social responsibility on company performance. IOP Conf Ser: Earth Environ Sci 235(1):012004

Afshari H, Farel R, Peng Q (2016) Need for optimization under uncertainty: designing flow exchanges in Eco-Industrial Parks. In: ASME 2016 international design engineering technical conferences and computers and information in engineering conference, V004T05A046–V004T05A046, American Society of Mechanical Engineers

Allenby BR (1992) Achieving sustainable development through industrial ecology. Int Environ Aff 4(1):56–68

Ayres RU, Ayres LW (eds) (2002) A handbook of industrial ecology. Edward Elgar Publishing, Northampton

Azapagic A, Clift R (1999) The application of life cycle assessment to process optimisation. Comput Chem Eng 23(10):1509–1526

Bain A, Shenoy M, Ashton W, Chertow M (2010) Industrial symbiosis and waste recovery in an Indian industrial area. Resour Conser Recycl 54(12):1278–1287

Bichraoui N, Guillaume B, Halog A (2013) Agent-based modelling simulation for the development of an industrial symbiosis-preliminary results. Procedia Environ Sci 17:195–204

Boix M, Montastruc L, Azzaro-Pantel C, Domenech S (2015) Optimization methods applied to the design of eco-industrial parks: a literature review. J Clean Prod 87:303–317

Bringezu S, Moriguchi Y (2002) Material flow analysis. In: A handbook of industrial ecology, pp 79–110

Bringezu S, Moriguchi Y (2018) Material flow analysis. In: Bartelmus P, Seifert EK (eds) Green accounting. Routeledge (2018).

Bringezu S, Fischer-Kowalski M, Kleijn R, Palm V (1997) Regional and national material flow accounting: From paradigm to practice of sustainability. In: Proceedings of the ConAccount workshop. Leiden, The Netherlands, 21–23 Jan 1997

Chertow MR (2000) Industrial symbiosis: literature and taxonomy. Ann Rev Energy Environ 25(1):313–337

Chew IML, Tan RR, Foo DCY, Chiu ASF (2009) Game theory approach to the analysis of inter-plant water integration in an eco-industrial park. J Clean Prod 17(18):1611–1619

Côté RP, Cohen-Rosenthal E (1998) Designing eco-industrial parks: a synthesis of some experiences. J Clean Prod 6(3):181–188

Côté R, Hall J (1995) Industrial parks as ecosystems. J Clean Prod 3(1):41–46

Curran T, Williams IDA (2012) Zero waste vision for industrial networks in Europe. J Hazard Mater 207:3–7

Domenech T, Davies M (2011) Structure and morphology of industrial symbiosis networks: the case of Kalundborg. Procedia-Soc Behav Sci 10:79–89

Drexhage J, Murphy D (2012) Sustainable development: from Brundtland to Rio 2012. In: Background paper for the high level panel on global sustainability. United Nations, New York

Duchin F (1992) Industrial input-output analysis: implications for industrial ecology. Proc Natl Acad Sci 89(3):851–855

Ehrenfeld JR (1997) Industrial ecology: a framework for product and process design. J Clean Prod 5(1):87–95

Ehrenfeld JR, Chertow MR (2002) Industrial symbiosis: the legacy of Kalundborg. In: Ayres RU, Ayres LW (eds) A handbook of industrial ecology. Edward Elgar

Erol P, Thöming J (2005) ECO-design of reuse and recycling networks by multi-objective optimization. J Clean Prod 13(15):1492–1503

Frosch RA (1994) Industrial ecology: minimizing the impact of industrial waste. Phys Today 47(11):63–68

Frosch RA, Gallopoulos NE (1989) Strategies for manufacturing. Sci Am 261(3):144–152

Geng Y, Fujita T, Park HS, Chiu ASF, Huisingh D (2016) Recent progress on innovative eco-industrial development. J Clean Prod 114:1–10

Gonela V, Zhang J (2014) Design of the optimal industrial symbiosis system to improve bioethanol production. J Clean Prod 64:513–534
Gu C, Estel L, Yassine A, Leveneur S (2013) Multi-objective optimization for industrial ecology: design and optimize exchange flows in an industrial park. In: Proceedings of the 2013 international conference on applied mathematics and computational methods (AMCM 2013), pp 109–116
Huang JJ, Chen CY, Liu HH, Tzeng GH (2010) A multi-objective programming model for partner selection-perspectives of objective synergies and resource allocations. Expert Syst Appl 37(5):3530–3536
Huo CH, Chai LH (2008) Physical principles and simulations on the structural evolution of Eco-Industrial systems. J Clean Prod 16(18):1995–2005
Jelinski LW, Graedel TE, Laudise RA, McCall DW, Patel CK (1992) Industrial ecology: concepts and approaches. Proc Natl Acad Sci 89(3):793–797
Karlsson M, Wolf A (2007) Using an optimization model to evaluate the economic benefits of industrial symbiosis in the forest industry. J Clean Prod 16(14):1536–1544
Kastner CA, Lau R, Kraft M (2015) Quantitative tools for cultivating symbiosis in industrial parks: a literature review. Appl Energy 155:599–612
Kuznetsova E, Louhichi R, Zio E, Farel R (2017) Input-output Inoperability Model for the risk analysis of eco-industrial parks. J Clean Prod 164:779–792
Lee S, Yoo C, Choi SK, Chun HD, Lee I-B (2006) Modeling of Eco-industrial Park (EIP) through material flow analysis (MFA). Korean Chem Eng Res 44:579–587
Li C, Zhang X, Zhang S, Suzuki K (2009) Environmentally conscious design of chemical processes and products: multi-optimization method. Chem Eng Res Des 87(2):233–243
Lowe EA (2001) Eco-industrial park handbook for Asian developing countries. Report to Asian Development Bank. Asian Development Bank. Oakland, CA
Mantese GC, Amaral DC (2017) Comparison of industrial symbiosis indicators through agent-based modeling. J Clean Prod 140:1652–1671
Miettinen K (1999) Nonlinear multiobjective optimization, vol 12. Springer
Pan AC (1998) Allocation of order quantity among suppliers. J Purch Mater Manage 25(3):36
Pishvaee MS, Jolai F, Razmi J (2009) A stochastic optimization model for integrated forward/reverse logistics network design. J Manuf Syst 28(4):107–114
Qu Y, Bektaş T, Bennell J (2014) Sustainability SI: multimode multicommodity network design model for intermodal freight transportation with transfer and emission costs. Netw Spat Econ 1–27
Roberts BH (2004) The application of industrial ecology principles and planning guidelines for the development of eco-industrial parks: an Australian case study. J Clean Prod 12(8):997–1010
Romero E, Ruiz MC (2013) Framework for applying a complex adaptive system approach to model the operation of eco-industrial parks. J Ind Ecol 17(5):731–741
Romero E, Ruiz MC (2014) Proposal of an agent-based analytical model to convert industrial areas in industrial eco-systems. Sci Total Environ 468:394–405
Schlarb M (2001) Eco-industrial development: a strategy for building sustainable communities. Rev Econ Dev Inter Pract 8
Taskhiri MS, Tan RR, Chiu AS (2011) Emergy-based fuzzy optimization approach for water reuse in an eco-industrial park. Resour Conserv Recycl 55(7):730–737
Theo W, Lim LS, Wan Alwi SA, Mohammad Rozali NE, Ho WSH, Abdul-Manan Z (2016) An MILP model for cost-optimal planning of an on-grid hybrid power system for an eco-industrial park. Energy 116:1423–1441
Tong L, Liu X, Liu X, Yuan Z, Zhang Q (2013) Life cycle assessment of water reuse systems in an industrial park. J Environ Manage 129:471–478
Tudor T, Adam E, Bates M (2007) Drivers and limitations for the successful development and functioning of EIPs (eco-industrial parks): a literature review. Ecol Econ 61(2):199–207
Veiga LB, Magrini A (2009) Eco-industrial park development in Rio de Janeiro, Brazil: a tool for sustainable development. J Clean Prod 17(7):653–661

Walther G, Schmid E, Spengler TS (2008) Negotiation-based coordination in product recovery networks. Int J Prod Econ 111(2):334–350

Yin Y (2002) Multiobjective bilevel optimization for transportation planning and management problems. J Adv Transp 36(1):93–105

Yuan Z, Bi J, Moriguichi Y (2006) The circular economy: a new development strategy in China. J Ind Ecol 10(1–2):4–8

Zitzler E, Thiele L (1999) Multiobjective evolutionary algorithms: a comparative case study and the strength Pareto approach. IEEE Trans Evol Comput 3(4):257–271

15

Prediction Models for Surface Roughness: A Comparative Study

Girish Kant Garg, Shailendra Pawanr, and Kuldip Singh Sangwan

Abstract Surface roughness is one of the significant index to measure the product quality of the machined parts. The objective of this work is to contribute towards the development of prediction models for surface roughness. In this work, the predictive models were developed for turning operations using soft computing techniques; support vector regression (SVR) and artificial neural network (ANN). The turning experiments are conducted to obtain the experimental data. The developed predictive models were compared using relative error and validated using hypothesis testing. The results indicate that both techniques provide a close relation between the predicted values and the experimental values for surface roughness and are appropriate to predict the surface roughness with significant acceptable accuracy. It is found that ANN performs better as compared to SVR.

Keywords Surface roughness · Artificial neural network · Support vector regression

15.1 Introduction

Predictive modelling is widely used in machining operations to improve the product quality, minimize the production cost and lower the power consumption. Surface roughness is one of the common index to measure the product quality of the machined parts (Sarikaya and Güllü 2014). Surface roughness of machined parts improves the fatigue strength, successive machining benefits, tribological characteristics, quality of fit in two mating parts and corrosive resistance etc. (Kant and Sangwan 2014). Literature depicted that surface roughness is one of the primary performance evaluation criterion in machining operations followed by machining cost and material removal rate (Yusup et al. 2012). The surface roughness of the machined parts depends upon various factors such as cutting parameters, properties of the work piece, cutting tool

G. K. Garg (✉) · S. Pawanr · K. S. Sangwan
Department of Mechanical Engineering, Birla Institute of Technology and Science, Pilani-Campus, Pilani 333031, Rajasthan, India
e-mail: girish@pilani.bits-pilani.ac.in

type and their geometry etc. The theoretical models available in the text books are unable to incorporate the dynamic behavior of the machining operations and failed to predict the surface roughness precisely (Davim et al. 2008). The researchers have used various modelling techniques to overcome the dynamic behavior of machining operations (Garg et al. 2016; Sangwan et al. 2015; Beatrice et al. 2014; Kant and Sangwan 2015a, b, c; Kant et al. 2013; Sangwan and Kant 2017; Pawanr and Garg 2019; Pal and Chakraborty 2005; Aykut Arapo˘glu and Mehmet Alper Sofuo˘glu 2017). Therefore, it is essential to accumulate and analyze the real time experimental data related to surface roughness and control parameters to develop a precise predictive model. In the present study, predictive model for the product quality in terms of surface roughness has been developed using support vector regression (SVR) and artificial neural network (ANN) during turning of mild steel 1045. Sandvik made carbide inserts were used as cutting inserts in a dry cutting environment for the turning operation. The proposed models were compared on the basis of relative error and validated using hypothesis testing.

15.2 Experimental Work

The main objective of this work is to develop predictive models due to which only three machining parameters are considered to simplify the modelling procedure. The turning parameters including cutting speed (v), feed rate (f) and cutting depth (d) and their levels are shown in Table 15.1.

Experiments were conducted on a heavy-duty HMT Centre lathe machine tool having maximum 2300 rpm and 5.5 kW motor rating. Mild steel grade AISI 1045 was selected as a work piece material for the turning because it has wide range of industrial and commercial applications. The carbide cutting inserts of grade TNMG 16 04 04 were used for cutting with tool holder PTGNR 2020K16. The Taylor and Hobson make profilometer was used to acquire the surface roughness data from the work piece. The surface roughness readings were taken on three equally spaced locations and their mean was computed to obtain the average surface roughness. The detailed information of experimental setup, measurements and design of experiment can be seen in reference Kant and Sangwan (2014).

Table 15.1 Machining parameters and their levels

Factor/levels	I	II	III
v (m/min)	103.31	134.30	174.14
f (mm/revolution)	0.12	0.16	0.2
d (mm)	0.5	1.0	1.5

15.3 Development of Predictive Models for Turning

In this section, the obtained experimental data is used to develop the predictive models using SVR and ANN.

15.3.1 Support Vector Regression

An online SVR toolbox for SVR modelling developed by Parrella (2007) in MATLAB is used for predicting the surface roughness. A combined vector of all the three input parameters (v, f, d) as a training set 'x' and the training set 'y' representing response parameter (surface roughness) is used. Twenty-seven sets of input-output pairs are used for training of the SVR model. The performance of the SVR model majorly depends upon the two variables known as insensitive loss function (ε) and cost function and are adjusted by the users to obtain the best outputs. Training parameters used for this study are $\varepsilon = 0.01$; $C = 1000$; kernel type = radial basis function (RBF); kernel parameter = 30. SVR checks the verification of Karush–Kuhn–Tucker (KKT) conditions and simultaneously trains the data one by one by adding each sample to the function. If the KKT conditions are not verified then the sample is stabilized using the stabilization technique, else the sample is added. To optimize the values, the stabilization technique dynamically changes the SVR parameters of insensitive loss function and cost function.

15.3.2 Artificial Neural Network

After several trials, it was found that the parameters shown in Table 15.2 leads to accurate results in minimum time.

The network structure 3-7-1 means that it consists of three neurons in the input layer, seven neurons in the hidden layer and output layer with one neuron. 22 random values from the experimental data were chosen for training and five random values were used for testing the network based on selected 80%:20% ratio. A feed forward back propagation algorithm was used to train the network by assigning random

Table 15.2 Selected ANN parameters for surface roughness prediction

ANN parameters	Value
Structure of network	3-7-1
Training/testing data	22/5
Performance function	Mean Square Error
Network algorithm	Feed forward back propagation
Transfer function	Logsig, tansig

Table 15.3 The weights and biases between neurons of input and hidden layer

Wjk, j = 3, k = 7	W1k	W2k	W3k	Bias (bk)
1	2.924	0.014	4.534	−5.241
2	3.582	0.831	−4.220	−4.490
3	−2.837	4.284	1.151	3.088
4	3.467	−4.160	3.723	−0.839
5	−1.004	5.115	2.841	1.423
6	−5.034	2.821	1.429	−7.357
7	−3.223	−4.798	5.559	−3.962

Table 15.4 The weights and bias between neurons of hidden and output layer

Wk, z k = 7, z = 1	W1k
1	0.755
2	0.291
3	0.566
4	−0.346
5	0.643
6	1.633
7	−0.656

weights and biases to interconnected neurons. This algorithm works on the principal of gradient decent method and updates the weights and biases in each iteration until the minimum mean square error is achieved between the target values and training values. The final weights and biases between the input layer and hidden layer are shown in Table 15.3. The weights between hidden layer and output layer are shown in Table 15.4. The bias value between hidden and output layer is 0.491. The neural network was trained using the parameters listed in Table 15.2 and it was observed that mean square error decreased until 250 iterations and after this point it was steady. The training was stopped after 250 iterations and the developed neural network was tested using the random experimental values, which were not used for training process.

15.4 Comparison and Validation of Predictive Models

Equation (15.1) is used to compute the relative errors between the predicted values by the proposed models and their respective experimental values of the surface roughness and are graphically presented in Fig. 15.1.

$$\text{Relative Error } (\%) = \frac{|\text{Predicted value - Experimental value}|}{\text{Experimental Value}} * 100 \qquad (15.1)$$

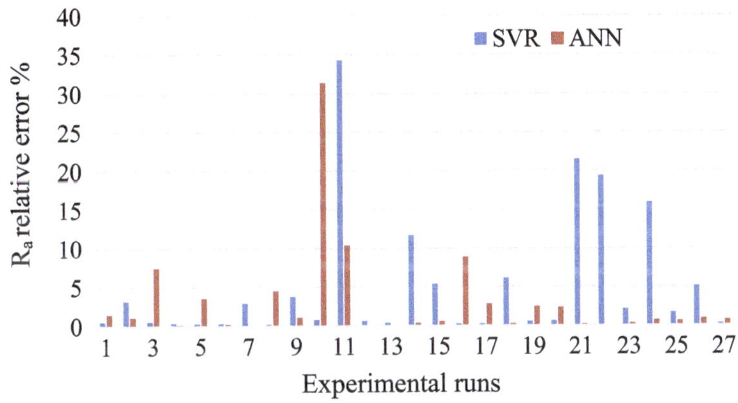

Fig. 15.1 Comparison of prediction capabilities of ANN and SVR in terms of relative error %

The average relative errors of 5.17% and 3.07% were found by SVR and ANN respectively.

Mean relative error illustrates that the ANN performs better as compared to SVR. It shows that the well-trained network model can take an optimal performance and has greater accuracy in predicting surface roughness as compared to SVR. Both the techniques are suitable for predicting the surface roughness in an acceptable range. However, the model generation and training procedure of ANN took more time as compared to SVR. Also, both the techniques are appropriate to predict the surface roughness with significant acceptable accuracy. Goodness of fit was calculated and compared for both techniques using some representative hypothetical tests and are shown in Table 15.5. These tests are *t*-test to test the means, *f*-test and Levene's test for variance. In all these tests, the *p*-values found to be greater than 0.05, which means that the null hypothesis cannot be rejected. The *p*-values in Table 15.5 also indicate that there is no significant evidence to conclude that the experimental data and the data predicted by SVR and ANN models differ. Therefore, both predictive models have statistically satisfactory goodness of fit for the modelling point of view.

Table 15.5 Results of Hypothesis testing to compare the models at 95% confidence level based on p-value

Tests	p-value	
	SVR	ANN
Mean paired t-test	0.840	0.882
Variance F-test	0.624	0.802
Levene's test	0.386	0.784

15.5 Conclusions

In this work, predictive models using soft computing techniques SVR and ANN were developed for surface roughness. SVR is capable of accurately predicting the surface roughness during turning operations. The mean relative error between the predicted and experimental values for the SVR model was found to be 5.17%. The predictive model developed using the ANN shows that the surface roughness values could be obtained with the selected ANN parameters. ANN has provided a close relation between the predicted values and the experimental values. The mean relative error for the predicted and experimental values using ANN was found to be 3.07%. It has been found that the model developed using ANN is capable of predicting accurately using a small number of training samples. The developed predictive models are compared using relative error and validated using hypothesis testing. It was found that ANN performs better as compared to SVR. The predictive capability could also be used for automatic monitoring. With the known boundaries of surface roughness and machining conditions, machining could be done with a relatively high rate of success leading to better surface finish.

References

Aykut Arapo˘glu R, Mehmet Alper Sofuo˘glu SO (2017) An ANN-based method to predict surface roughness in turning operations. Arab J Sci Eng 42:1929–1940. https://doi.org/10.1007/s13369-016-2385-y

Beatrice BA, Kirubakaran E, Thangaiah PRJ, Wins KLD (2014) Surface roughness prediction using artificial neural network in hard turning of AISI H13 steel with minimal cutting fluid application. Procedia Eng 97:205–211. https://doi.org/10.1016/j.proeng.2014.12.243

Davim JP, Gaitonde VN, Karnik SR (2008) Investigations into the effect of cutting conditions on surface roughness in turning of free machining steel by ANN models. J Mater Process Technol 205:16–23. https://doi.org/10.1016/j.jmatprotec.2007.11.082

Garg A, Lam JSL, Gao L (2016) Power consumption and tool life models for the production process. J Clean Prod 131:754–764. https://doi.org/10.1016/j.jclepro.2016.04.099

Kant G, Sangwan KS (2014) Prediction and optimization of machining parameters for minimizing power consumption and surface roughness in machining. J Clean Prod 83:151–164. https://doi.org/10.1016/j.jclepro.2014.07.073

Kant G, Sangwan KS (2015a) Predictive modelling and optimization of machining parameters to minimize surface roughness using artificial neural network coupled with genetic algorithm. Procedia CIRP 31:453–458. https://doi.org/10.1016/j.procir.2015.03.043

Kant G, Sangwan KS (2015b) Predictive modeling for power consumption in machining using artificial intelligence techniques. Procedia CIRP 26:403–407. https://doi.org/10.1016/j.procir.2014.07.072

Kant G, Sangwan KS (2015c) Predictive modelling for energy consumption in machining using artificial neural network. Procedia CIRP 37:205–210. https://doi.org/10.1016/j.procir.2015.08.081

Kant G, Rao VV, Sangwan KS (2013) Predictive modeling of turning operations using response surface methodology. Appl Mech Mater 307:170–173. https://doi.org/10.4028/www.scientific.net/AMM.307.170

Pal SK, Chakraborty D (2005) Surface roughness prediction in turning using artificial neural network. Neural Comput Appl 14:319–324. https://doi.org/10.1007/s00521-005-0468-x

Parrella F (2007) Online support vector regression. Thesis in Information Science, University of Genoa, Italy

Pawanr S, Garg GK (2019) Selection of optimum Cutting Parameters for Minimization of Specific Energy Consumption during Machining of Al 6061. J Phys: Conf Ser 1240:012064. https://doi.org/10.1088/1742-6596/1240/1/012064

Sangwan KS, Kant G (2017) Optimization of machining parameters for improving energy efficiency using integrated response surface methodology and genetic algorithm approach. Procedia CIRP 61:517–522. https://doi.org/10.1016/j.procir.2016.11.162

Sangwan KS, Saxena S, Kant G (2015) Optimization of machining parameters to minimize surface roughness using integrated ANN-GA approach. Procedia CIRP 29:305–310. https://doi.org/10.1016/j.procir.2015.02.002

Sarikaya M, Güllü A (2014) Taguchi design and response surface methodology based analysis of machining parameters in CNC turning under MQL. J Clean Prod 65:604–616. https://doi.org/10.1016/j.jclepro.2013.08.040

Yusup N, Mohd A, Zaiton S, Hashim M (2012) Expert systems with applications evolutionary techniques in optimizing machining parameters: review and recent applications (2007–2011). Expert Syst Appl 39:9909–9927. https://doi.org/10.1016/j.eswa.2012.02.109

16

Solar Energy Forecasting Model and its Development

Sumanta Pasari and Venkata Satya Siva Koundinya Nandigama

Abstract Renewable energy comprises solar, wind, tidal, biomass and geothermal energies. Use of renewable energy resources as a substitute for fossil fuels inevitably reduce environmental footprint. Therefore, integration of renewable energy to the power grid, smart grid planning and grid-storage preparations are some of the major concerns in all developing countries. However, unpredictability in renewable energy resources makes the situation challenging. In light of this, the present study aims to develop a solar energy forecasting model to estimate future energy supply for a smooth integration of solar energy to the current electric grids. A suite of eight probability models, namely exponential, gamma, normal, lognormal, logistic, log-logistic, Rayleigh and Weibull distributions are used. While the model parameters are estimated from the maximum likelihood estimation method, the performance of the candidate distributions is tested using three goodness of fit tests: Akaike information criterion, Chi-square criterion, and K-S minimum distance criterion. Based on the sample data obtained from the Charanka Solar Park, Gujarat, it is observed that the Weibull model provides the best representation to the observed solar radiations. The study concludes with the analysis of forecasted solar energy and its possible role in replacing thermal energy resources.

Keywords Solar energy · Probability distribution · Charanka solar park

16.1 Introduction

World energy resources can be broadly classified into three types: fossil fuels, renewable resources, and nuclear resources. Renewable energy is the energy collected from naturally replenished sources such as sunlight, wind, tides, biomass and geothermal

S. Pasari (✉) · V. S. S. K. Nandigama (Deceased)
Department of Mathematics, Birla Institute of Technology and Science, Pilani Campus, Pilani 333031, RJ, India
e-mail: sumanta.pasari@pilani.bits-pilani.ac.in; 86.sumanta@gmail.com

V. S. S. K. Nandigama (Deceased)
e-mail: f2013108p@alumni.bits-pilani.ac.in

heat. It contributed to almost 20% to human's global energy consumption and 25% to global electricity generation in 2015 and 2016, respectively (REN 21 homepage 2019; Global energy homepage 2019). India is one of the largest renewable energy producing countries accounting for about 35% of the total installed power capacity in the electricity sector. The target by 2030, as stated in the Paris Agreement, is to achieve 40% of total India's electricity generation from non-fossil fuel sources (Global energy homepage 2019; Paris Agreement homepage 2019; Rather 2018). As a consequence, a large number of wind and solar energy plants are being installed in the country under the purview of the Ministry of New and Renewable Energy (MNRE) (Paris Agreement homepage 2019; Rather 2018).

Despite the installation of many renewable energy plants, their integration to the main power grids is crucial in harnessing renewable energy applications (REN 21 homepage 2019; Global energy homepage 2019; Paris Agreement homepage 2019; Rather 2018; Zhang et al. 2015). The unpredictability of renewable energy resources, such as wind speed and solar radiation makes integration difficult, as the current electric grids cannot operate unless there is a mutual balance between supply and demand (Zhang et al. 2015; Su et al. 2012; Jacobson and Delucchi 2011; Delucchi and Jacobson 2011; NREL homepage 2019). An imbalance may result in voltage fluctuations and even worse (NREL homepage. https://www.nrel.gov/. 2019). Other problems related to renewable energy sources include the unavailability of solar power at night during which the power consumption is at its peak and the lack of efficient energy storage systems to save the excess electricity production (Delucchi and Jacobson 2011; NREL homepage 2019). In addition, as renewable energy plants are usually located far away from the consumption location, transportation of power may cause unwanted transmission losses (Zhang et al. 2015; Su et al. 2012; Jacobson and Delucchi 2011; Delucchi and Jacobson 2011; NREL homepage 2019).

Several methods are employed for the forecasting of solar irradiation considering numerical weather prediction, artificial neural networks (ANN), linear and non-linear stochastic models, remote sensing based models and hybrid models (Ferrari et al. 2013; Zhang et al. 2015; Inman et al. 2013). Comparison of several autoregressive models (AR, ARMA, ARIMA) (Ferrari et al. 2013) and neural network based models such as Radial Basis Function Neural Networks (RBFNN), Least Square Support Vector Machine (LS-SVM), k-Nearest Neighbour (kNN), and Weighted kNN (WkNN) methods (Zhang et al. 2015) have been implemented as forecasting engines. Use of empirical probability models (Pasari 2015, 2018) could also be tried for energy forecasting. In summary, two main categories of studies have evolved, one focusing on the smart grid or grid energy storage technology and another aiming at forecasting of renewable energy (Rather 2018; Zhang et al. 2015; Su et al. 2012; Jacobson and Delucchi 2011; Delucchi and Jacobson 2011; NREL homepage 2019). The present study considers the latter issue and concentrates on the statistical modeling of solar power output at Charanka Solar Park, Gujarat. The aim is to select the best-fit probability distribution(s) among exponential, gamma, normal, lognormal, logistic, log-logistic, Rayleigh and Weibull models to forecast solar radiations.

16.2 Data Description

Solar radiation, the radiant energy emitted by the sun, is the primary data for the present analysis. When solar radiation enters into the Earth's atmosphere, a fraction of the radiation reaches directly to the surface. Such radiation is called beam or direct radiation. The remaining fraction may be scattered or absorbed by air molecules, clouds or aerosols. A part of such scattered radiation reaches the ground and is known as diffuse radiation. Another part of the direct radiation hitting the surface gets reflected and may reach upon another surface, such as solar collector or photovoltaic panel. Such radiation is called albedo. The sum of these three components is termed as global radiation (Rather 2018). The quantum of global irradiation collected per unit area is an important parameter for solar power forecast.

Direct Normal Irradiance (DNI) is the amount of solar radiation received per unit area by a surface that is always held perpendicular to the rays coming in a straight line from the direction of the sun at its current position in the sky. Diffuse Horizontal Irradiance (DHI), on the contrary, is the amount of radiation received per unit area by a surface that does not arrive on a direct path from the sun, but has been scattered by molecules and particles in the atmosphere and comes equally from all directions. Global Horizontal Irradiance (GHI) is the total amount of shortwave radiation received from above by a surface horizontal to the ground (Rather 2018). The GHI may be calculated from DNI and DHI as

$$GHI = DHI + DNI * \cos(\theta) \tag{16.1}$$

Where θ is the solar zenith angle (Rather 2018; Zhang et al. 2015).

The data of the Charanka Solar Power Park (23.95° N, 71.15° E) in Gujarat was procured from the National Solar Radiation Database of National Renewable Energy Laboratory (NREL) (NREL homepage 2019). It comprises hourly data of all the variables (e.g., DNI, DHI, GHI, and many others) affecting the solar irradiation from 2000 to 2014. It is observed that depending on the season, about 12 h of daily solar irradiation data (06:30–18:30 h) contain non-zero positive entries of DNI, DHI and GHI values. There are many zero values in the sample data indicating that the day did not start or the day had ended. To maintain consistency in the analysis, 08 h of daily data (09:30–16:30 h) is considered for modeling. With this filtering, yearly 2920 data points are obtained. All DNI, DHI and GHI data are fitted separately to identify the best-fit probability model(s) for solar power forecast.

It may be noted that the original datasets also contain information on temperature, pressure, relative humidity and precipitation among others, although those are not used in the present analysis.

16.3 Methodology and Results

On a temporal scale, solar power forecasting may be classified into now casting (forecasting up to a few hours in advance), short-term forecasting (forecasting up to a few days in advance) and long-term forecasting (forecasting months or years ahead) (Rather 2018; Zhang et al. 2015; Su et al. 2012; Jacobson and Delucchi 2011; Delucchi and Jacobson 2011). Depending upon the range of forecasts required, forecasting models have been developed accordingly incorporating parameters that are affecting solar radiation in the range (Zhang et al. 2015). Both short and long-term power forecasts have their specific applications. While system operators use short term forecasts in unit commitment analysis and determining reserve unit requirements, solar farm owners use such forecast for bidding strategy planning (in electricity markets) and dealing with voltage imbalance issues while integrating solar power supply to major thermal power distribution networks (Rather 2018; Zhang et al. 2015; Su et al. 2012; Jacobson and Delucchi 2011; Delucchi and Jacobson 2011; NREL homepage 2019). The long-term solar power forecasts are particularly important for smart city planning and negotiating contracts with financial entities or utilities (Zhang et al. 2015). Statistical approaches, as in this study, are preferred for long-term forecasts.

The methodology here comprises three major steps: probability model assumption, parameter estimation, and model validation. Based on some graphical representation of data, eight probability models are considered to fit DNI, DHI and GHI data separately.

Model parameters of the studied distributions are estimated from the classical maximum likelihood estimation (MLE) method, whereas the model selection is performed based on three goodness of fit tests, namely Akaike information criterion (AIC), Chi-square criterion and K-S minimum distance criterion. The AIC test is a simple modification from the log-likelihood scores and it accounts for the additional number of parameters in the competitive models.

The Kolmogorov–Smirnov (K–S) test, in contrast, is a non-parametric approach. The Chi-square test determines significant differences between the expected and observed frequencies in one or more categories (Ferrari et al. 2013; Zhang et al. 2015). The results of estimated parameters and selection scores corresponding to average GHI data are presented in Table 16.1.

It may be noted that each parameter in the studied distributions has its respective role (e.g., shape, scale, and location) (Pasari 2015, 2018). Moreover, like the results in Table 16.1 for GHI data, one may obtain results corresponding to DHI and DNI data using simple excel tools along with Matlab plots. It is observed that the Weibull model consistently provides the best representation for DHI, DNI and GHI data. The pictorial representation of the model fit for DHI, DNI and GHI data of year 2000 is illustrated in Figs. 16.1, 16.2 and 16.3.

With the above process of finding the best fit probability distribution, one can now analyze solar irradiation data for future estimation. As a secondary illustration, forecasting of solar irradiance may be carried out using a simple linear regression model (Rather 2018; Zhang et al. 2015; Su et al. 2012; Jacobson and Delucchi

Table 16.1 Estimated parameter values and model selection results for the GHI data

Distribution	Parameter values		Model selection		
			AIC	Chi-Square	K–S
Exponential	$\hat{\lambda}$ (scale)	468.01	334927.5	38992.2	0.3601
Gamma	$\hat{\alpha}$ (scale)	340.61	298718.7	2311.84	0.0757
	$\hat{\beta}$ (shape)	1.37			
Normal	$\hat{\mu}$ (location)	468.01	295710.6	421.84	0.0413
	$\hat{\sigma}$ (scale)	272.73			
Lognormal	$\hat{\mu}$ (log-scale)	5.74	301470.0	4349.08	0.0902
	$\hat{\sigma}$ (shape)	1.27			
Logistic	$\hat{\mu}$ (location)	476.86	296652.3	695.22	0.0467
	$\hat{s}$ (scale)	165.17			
Log-logistic	$\hat{\alpha}$ (scale)	5.98	299742.4	3751.30	0.0871
	$\hat{\beta}$ (shape)	0.59			
Rayleigh	$\hat{\alpha}$ (scale)	383.01	309226.2	10596.11	0.2297
Weibull	$\hat{\alpha}$ (scale)	505.43	295072.0	360.61	0.0348
	$\hat{\beta}$ (shape)	1.44			

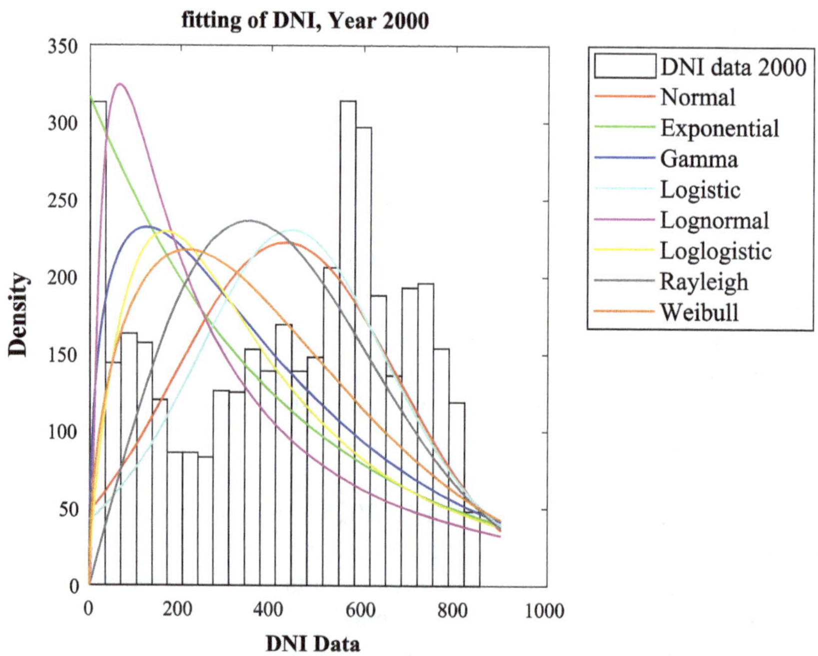

Fig. 16.1 Data fit of DNI values for the year 2000

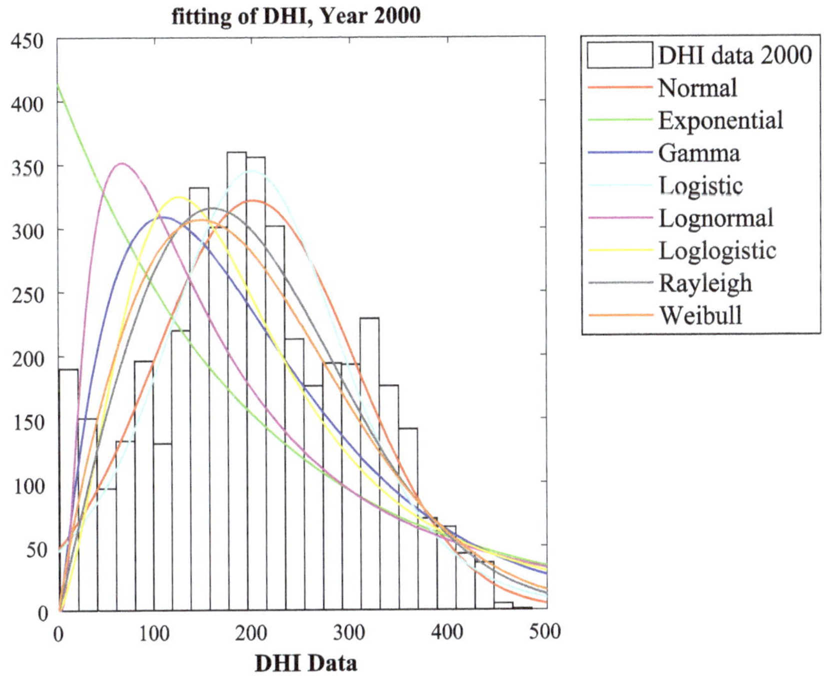

Fig. 16.2 Data fit of DHI values for the year 2000

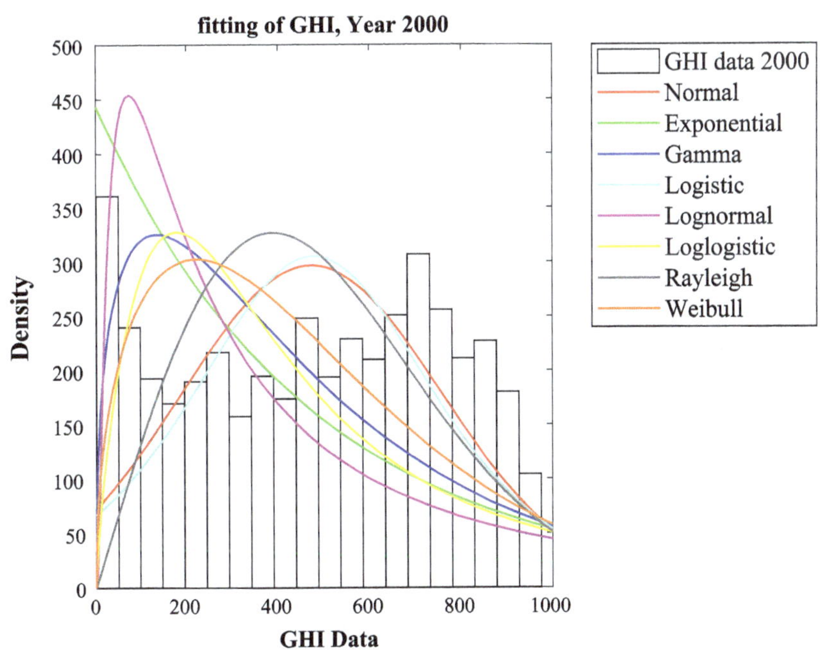

Fig. 16.3 Data fit of GHI values for the year 2000

Table 16.2 Regression analysis of total DHI during 2000–2014 for the month of March

Summary output: regression statistics		
	Coefficients	Standard error
Multiple R	0.2294	
R Square	0.0526	
Adjusted R-Square	0.0506	
Standard Error	245.8266	
Observations	465	
Multiple R	0.2294	
Intercept	2292.7576	22.8367
Variable	0.4306	0.08493

2011; Delucchi and Jacobson 2011). The regression analysis can well explain the relationship between criterion variables (dependent variables) and predictor variables (independent variables). Then those values are interpolated or extrapolated with the help of the relationship obtained by the regression model. In the present study, the analysis is performed with the help of MS-excel inbuilt statistical data analysis tool. The data is forecasted for 5 additional years from 2015 to 2019 month wise from the results of the linear regression analysis based on 2000–2014. Also, the solar irradiation is forecasted using the first 10 years of the sample data from 2000 to 2009 and the corresponding errors have been calculated. For a demonstration, the regression analysis for the month of March is provided in Table 16.2 below. Table 16.2 shows that the adjusted R-squared value is 0.0506, while the multiple R value is 0.2294. Similarly, regression analysis can be easily performed for all the months of the calendar. However, improvements in the results are necessary.

16.4 Summary and Conclusions

Statistical modeling of renewable energy plays a pivotal role in the future energy sector and therefore its importance can never be disregarded. In this work, first the best-fit probability model of solar irradiation data is identified using eight popular probability distributions. Then a linear regression analysis is carried out to forecast solar energy for the Charanka Solar Park, Gujarat. During the course of the study, the following important observations are noted:

- As the day progresses, the amount of DHI, DNI and GHI values increases till afternoon and then decreases. This is because the zenith angle gradually decreases to zero as the day advances to afternoon and then the zenith angle gradually increases as the day advances into evening followed by night.
- The best-fit distribution for a particular hour over the months remains consistent although with varying means. This may be attributed to the variation in the amount of solar irradiation received on account of the seasons in respective years.

- The standard deviations of the fitted distribution are very high. Even the MSE (Mean Squared Error) for the regression is also high probably due to the less amount of data points.

As a conclusion, the present work has provided a layout to develop a solar energy-forecasting model towards the endeavor of estimating future energy supply for a smooth integration of solar energy to the current electric grids. Results, based on the limited data, are preliminary and require further analysis for a stringent conclusion.

References

Delucchi MA, Jacobson MZ (2011) Providing all global energy with wind, water, and solar power, part II: reliability, system and transmission costs, and policies. Energy Policy 39(3):1170–1190. https://doi.org/10.1016/j.enpol.2010.11.045

Ferrari S, Lazzaroni M, Piuri V, Loredana C, Marco F (2013) Statistical models approach for solar radiation prediction. In: 2013 IEEE international instrumentation and measurement technology conference (I2MTC), pp 1734–1739. https://doi.org/10.1109/i2mtc.2013.6555712

Global energy homepage. https://www.iea.org. Last accessed 24 June 2019

REN 21 homepage. http://www.ren21.net/gsr-2017/. Last accessed 24 June 2019

Inman RH, Pedro HTC, Coimbra CFM (2013) Solar forecasting methods for renewable energy integration. Prog Energy Combust Sci 39(6):535–576. https://doi.org/10.1016/j.pecs.2013.06.002

Jacobson MZ, Delucchi MA (2011) Providing all global energy with wind, water, and solar power, part I: technologies, energy resources, quantities and areas of infrastructure, and materials. Energy Policy 39(3):1154–1169. https://doi.org/10.1016/j.enpol.2010.11.040

NREL homepage. https://www.nrel.gov/. Last accessed 24 June 2019

Paris Agreement homepage. https://www4.unfccc.int/sites/submissions/indc/Submission%20Pages/submissions.aspx. Last accessed 24 June 2019

Pasari S (2015) Understanding himalayan tectonics from geodetic and stochastic modeling. PhD dissertation. Indian Institute of Technology Kanpur, India

Pasari S (2018) Stochastic modeling of earthquake interoccurrence times in northwest Himalaya and adjoining regions. Geomatics, Nat Hazards Risk 9(1):568–588. https://doi.org/10.1080/19475705.2018.1466730

Rather NUR (2018) Introduction to renewable energy technologies in India. Education Publishing, pp 1–140

Su Y, Chan LC, Shu L, Tsui KL (2012) Real-time prediction models for output power and efficiency of grid-connected solar photovoltaic systems. Appl Energy 93:319–326. https://doi.org/10.1016/j.apenergy.2011.12.052

Zhang Y, Beaudin M, Taheri R, Zareipour H, Wood D (2015a) Day-ahead power output forecasting for small-scale solar photovoltaic electricity generators. IEEE Trans Smart Grid 6(5):2253–2262. https://doi.org/10.1109/tsg.2015.2397003

Zhang Y, Beaudin M, Taheri R, Zareipour H, Wood D (2015) Day ahead power output forecasting for small-scale solar photovoltaic electricity generators. IEEE Trans Smart Grid 6(5):2253–2262. https://doi.org/10.1109/tsg.2015.2397003

17

Ceramic Sanitary Ware Supply Chain: Assessing Sustainability

Kuldip Singh Sangwan, Kailash Choudhary, and Shilpi Agarwal

Abstract Sanitary wares are the integral part of construction materials but there is hardly any study in the literature which shows the environmental impacts from the sanitary ware. This paper aims at assessing sustainability of a ceramic sanitary ware supply chain by quantifying the environmental impacts from materials and resources used throughout the different phases of a sanitary ware life cycle. The impacts are quantified using ReCiPe endpoint and midpoint assessment methods with Umberto NXT Software and eco-invent 3.0 database. This study uses climate change, fossil depletion, human toxicity, metal depletion, ozone depletion, terrestrial acidification, water depletion, damage to ecosystem quality, human health, and resources assessment categories to quantify the environmental impacts. The life cycle assessment finds that consumption of heavy fuel oil, electricity, grass, and cement mortar is primarily responsible for the negative impacts on the environment. It is also found that manufacturing and transportation phases of the supply chain have maximum contribution to the environmental degradation. The methodology, assessment methods and impact categories used in the study can be used by the other ceramic enterprises for the identification and benchmarking of environmental hotspots in their supply chains. It is expected that this study will be useful for the policy makers as well as the manufacturer to find the key areas for decreasing the environmental impacts and enhancing sustainability of a sanitary ware supply chain.

Keywords Life cycle assessment · Sustainability assessment · Ceramic sanitary ware

---

K. S. Sangwan (✉) · S. Agarwal
Birla Institute of Technology and Science, Pilani 333031, India
e-mail: kss@pilani.bits-pilani.ac.in

K. Choudhary
Mody University of Science and Technology, Lakshmangarh, India

17.1 Introduction

Sustainable construction has started getting researchers' and policy makers' attention because of high amount of energy and resource consumption by construction sector. The construction sector is responsible for the 40% energy and 30% raw material consumption worldwide. The construction sector is responsible for the 25% of the world's solid waste generation and the ceramic waste contributes highest to this. The sanitary ware market is driven by the increasing standards of living, growing real estate industry, and increasing population in the developing countries. The ceramic sanitary ware market is estimated to record a compound annual growth rate (CAGR) of 10.3% between 2016 and 2026. The Indian sanitary ware market share accounts for 8% of the global production and stands at second rank in terms of volume in the Asia-Pacific region. In India, around 69% households in rural areas and about 19% households in urban areas still do not have access to safe sanitation (http://www.gima.de/media/1/1/4798/overview_of_india_s_bathroom_kitchen_industry.pdf). This provides an opportunity in term of sanitary ware market growth but it is also a cause of worry from the environmental perspective because of large quantity of natural resource extraction and waste generation along the supply chain of a sanitary ware.

The processes used in the production of ceramics are broadly divided into three groups; namely, pre kiln, kiln and post kiln processes. The major pre kiln processes contain ball milling, blunging, mixing, casting, drying, cleaning, and glazing. The pre kiln waste is called green waste, as this is recyclable without any difficulty. Generally, the green waste is 40% of the green product weight. The green products are then fired in the kiln to provide strength. Any rejection/waste after kiln firing (post kiln) is non-recyclable. The major problems faced by the ceramic industry are large rejections and poor quality. The large rejections, particularly after kiln firing, cause environmental issues and problems of land occupation. The current choice to dispose off the ceramic waste is landfill (Bhamu and Sangwan 2015).

Life cycle assessment (LCA) is a tool to assess, compare and benchmark the environmental impacts associated with pre-manufacturing, manufacturing and post-manufacturing phases of a product. Generally, the LCA studies are carried out with different scopes: cradle to cradle, cradle to grave, cradle to gate, and gate to gate. The LCA concept came into existence during 1970–90 but the standardization of approaches and terminologies took place during 1990–2000 (Guinee et al. 2011). There are a few papers in literature showing the environmental impact assessment from the construction materials, and construction & demolition (C&D) waste (Ye et al. 2018; Özkan et al. 2016; Sangwan et al. 2018) but none of these studies has focus on assessment of environmental impact due to sanitary ware products. Thus, the objective of this study is to assess and map the environmental impacts of different phases, processes, and materials used throughout the product life cycle including end of life disposal (pre-manufacturing, manufacturing, and post-manufacturing).

The next section of study presents the background of the selected topic. Section 17.3 presents the materials and methods used in the study. Section 17.4 shows the observed results and discussion. Section 17.5 explains the practical implications of the findings and the study ends with conclusions in Sect. 17.6.

17.2 Background

There are two schools of thought to assess the environmental impacts arising out during a product's life cycle. One school uses the knowledge of professionals to assess the environmental impacts using survey based empirical investigations (Mittal and Sangwan 2014). This is a simple method but does not provide the quantified environmental impacts generated during a product life cycle. The second school of thought provides quantitative emission results which can be easily attributed to the resources and materials used during different phases of a product life cycle (Ye et al. 2018; Özkan et al. 2016; Sangwan et al. 2018). This method is more exhaustive but requires lots of quantitative data for the assessment. The assessment can show a product's environmental impact in terms of climate change potential, freshwater ecotoxicity potential, freshwater eutrophication potential, marine ecotoxicity potential, marine eutrophication potential, terrestrial acidification potential, terrestrial eutrophication potential, fossil depletion potential, human toxicity potential, metal depletion potential, ozone depletion potential, particulate matter formation potential, terrestrial acidification potential, agriculture land occupation, urban land occupation, ionizing radiation, natural land transformation, photochemical oxidant formation, and water depletion potential. The possible assessment methods to calculate the environmental impacts using different LCA tools are: ReCiPe endpoint, ReCiPe midpoint, CML2001, impact2002+ , EDIP2003, and ecoindicator99 (Rebitzer et al. 2004). LCA provides the environmental hotspots along the supply chain, therefore remedial measures are much more effective to control the environmental impacts. This paper belongs to the second school of thought.

A few studies have been done to assess the environmental impacts generated during the production of construction materials. Ye et al. (2018) conducted a cradle to gate life cycle assessment and life cycle cost analysis for ceramic tile manufacturing in China. The results show that major environmental impacts are in terms of marine ecotoxicity, climate change, terrestrial ecotoxicity, human toxicity, and fossil depletion. The study observes that the use of inorganic chemicals and coal-based electricity are the main factors contributing to the generation of environmental impacts. Özkan et al. (2016) conducted a cradle to gate life cycle assessment study for the refractory bricks. The study found that the production of raw materials and firing process are the main contributors to the environmental impacts. Balasbaneh et al. (2018) conducted sustainability assessment to find the environmental friendly design alternate among the different options for wall and roof construction using

LCA. Sangwan et al. (2018) assesses the environmental impacts in terms of mid-point and endpoint categories during the different life cycle phases, from material and resources consumed along a ceramic tile supply chain. The study identifies that manufacturing phase generates highest environmental impacts due to the use of red oxide in glaze preparation, electricity consumed in kiln, and film used in packaging. Bribián et al. (2011) conducted an LCA of different construction materials (steel, cement, ceramics, mortar, wood, etc.). The study identified that ceramics materials in construction account for 21.5% of the energy consumption and 20.3% of CO_2 emissions. This is due to the electricity and natural gas consumption in the kiln process. Cabeza et al. (2014) presents the review of the studies related to life cycle assessment, life cycle energy analysis, and life cycle cost analysis for building and building related industries. The review shows that there are only few studies on the life cycle assessment of the construction materials, their production and selection.

The literature shows the importance of quantification and assessment of the environmental impacts from the different materials required in the construction industry. But there is hardly any study which shows the environmental impacts from the supply chain of ceramic sanitary wares. This study tries to fill this gap by quantifying and assessing the environmental impacts during the different life cycle phases, processes, resources, and materials in a sanitary ware supply chain.

17.3 Materials and Method

Environmental impacts from sanitary ware supply chain have been assessed using LCA framework based on ISO 14040 series guidelines using primary and secondary data (Rebitzer et al. 2004). The LCA framework uses four steps. In the first step; study goal, scope and functional unit are defined. This is followed by the collection of inventory data for the study. Next, impact assessment is done using LCA software followed by the interpretation of the results.

17.3.1 Goals and Scope

The goals of this study are to:

I. Assess the environmental impacts from a sanitary ware supply chain.
II. Compare the environmental impacts caused in different phases due to consumption of different materials and resources along the product life cycle.

A cradle to grave approach has been used, which comprises of raw material extraction, manufacturing, transportation, installation, and disposal phases (Rebitzer et al. 2004). Use phase is excluded due to the lack of data availability. The functional unit considered for this study is a standard washbasin weighing 14 kg. Machine life in manufacturing phase is assumed to be 20 years. Process flow in Fig. 17.1 shows

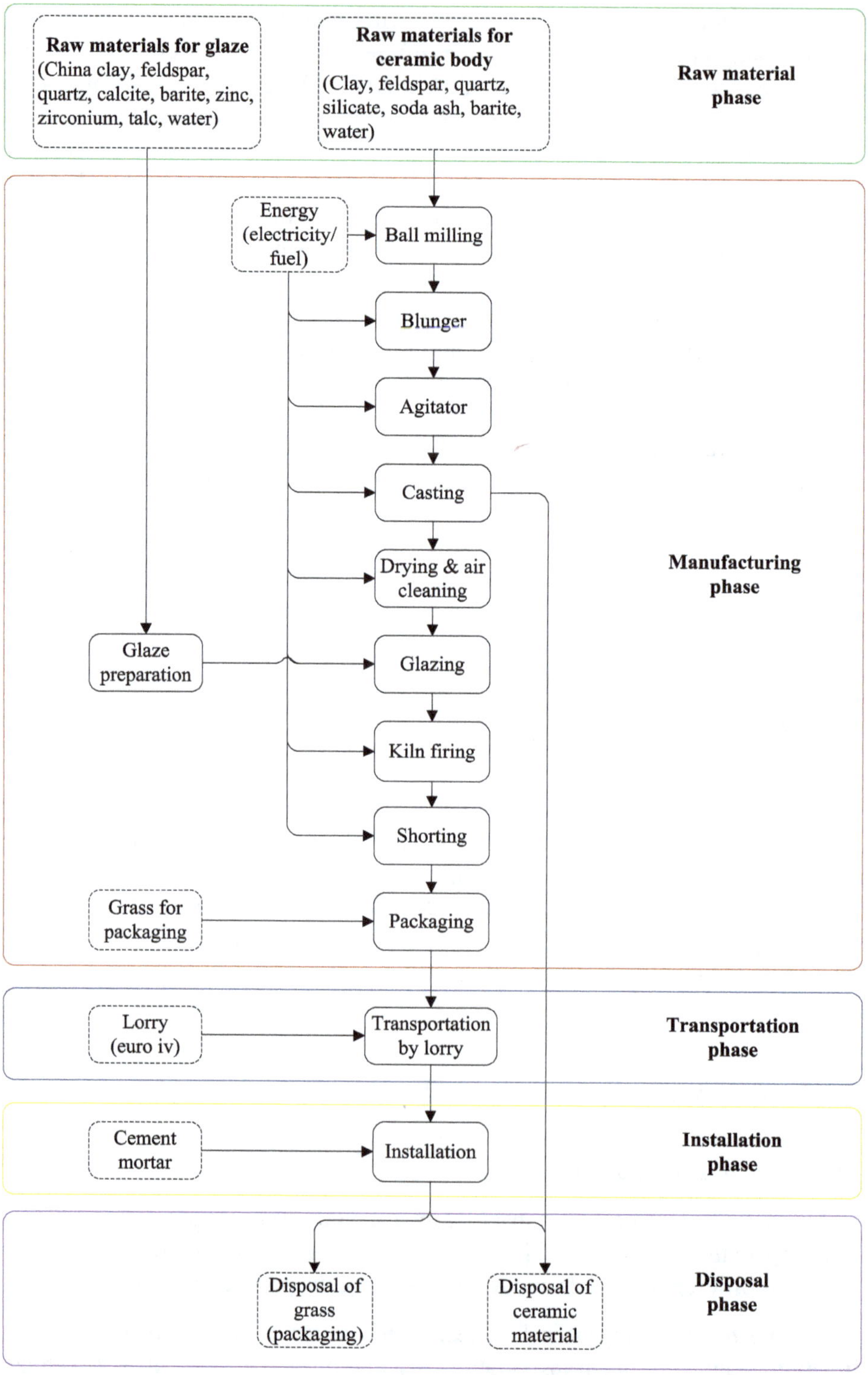

Fig. 17.1 Process flow of sanitary ware supply chain with different phases and processes

Table 17.1 Primary inventory data for one sanitary ware product (14 kg standard washbasin)

Processes	Materials/resource, energy, and emissions
Ball milling	Clay 9.1 kg, Feldspar 3.5 kg, Quartz 0.7 kg, Silicate 0.056 kg, Soda Ash 0.028 kg, Barite 0.007 kg, Water 6.3 l, electricity 0.088 kwh
Blunzer	Soda Ash 0.021 kg, Silicate 0.042 kg, Water 4.2 l, electricity 0.081 kwh
Casting (POP Mold)	Gypsum 0.4 kg, Water 0.21 l, Solid waste 0.61 kg
Glaze preparation	Feldspar 0.072 kg, Quartz 0.036 kg, Calcite 0.026 kg, China Clay 0.012 kg, Barite 0.004 kg, Zinc 0.006 kg, Zirconium 0.024 kg, Talc 0.001 kg, Water 0.072 l, electricity 0.001 kwh
Kiln firing	Heavy fuel oil 1.03 l, electricity 0.41 kwh
Packaging	Grass 0.92 kg
Installation	Mortar 4.75 kg, packaging waste 0.92 kg, Solid waste 14 kg (after use)
Total electricity	0.8 kwh (in all processes)

the scope of the study and major processes in this supply chain. The Umberto model is developed after carefully studying the processes and supply chain of a sanitary ware manufacturer. The developed model used for the calculations of environmental impacts contains the detailed information of processes, materials and resources used in the supply chain. Unlike the developed Umberto model, process flow diagram (Fig. 17.1) shows only major processes in the supply chain.

17.3.2 Inventory Analysis

A case study was conducted at one of the manufacturers of sanitary ware products located in industrial area of Bikaner city in India. This manufacturer, a medium scale enterprise, supplies sanitary ware to the northern part of India. The sanitary ware production process is standard. One of the authors studies the manufacturing process at the site to get the average of the inventory data. This primary data contains the quantity of raw materials (clay, feldspar, quartz, etc.), water, fuel, and electricity used along the supply chain as shown in Table 17.1. Secondary data for upstream activities (raw material extraction, electricity, fuel, machine consumption, and water supply) and downstream activities (ceramic waste disposal and packaging waste disposal) is taken from eco-invent 3.0 database.

17.3.3 Impact Assessment

Umberto NXT software with eco-invent 3.0 database is used for environmental impact assessment using ReCiPe endpoint and ReCiPe midpoint assessment methods. The ReCiPe method is selected from the other available methods of life cycle impact assessment (LCIA) because this is a newly developed method on the concepts of CML 2001 and eco-indicator 99. This method harmonizes the existing midpoint and endpoint assessment methods and eliminates the barrier of LCIA method selection in design and analysis of LCA as pointed out by Dong and Ng (2014). The impact categories considered in this study for ReCiPe endpoint assessment are damage to ecosystem quality, human health, and resources; and climate change (kg CO_2-Eq), fossil depletion (kg oil-Eq), human toxicity (kg 1,4-DCB-Eq), metal depletion (kg Fe-Eq), ozone depletion (kg CFC-11-Eq), terrestrial acidification (kg SO_2-Eq), and water depletion (m^3) for ReCiPe midpoint assessment.

17.4 Results and Discussion

The results of the environmental impacts from materials and resources used throughout the different phases of a sanitary ware life cycle are presented here.

17.4.1 Endpoint Assessment

The Fig. 17.2 show the environmental impact assessment using ReCiPe endpoint assessment method. It depicts the impact from various phases (i.e. raw material extraction, manufacturing, transportation, installation, and disposal) in the categories of ecosystem quality, human health, and resources. As shown in Fig. 17.2, manufacturing phase has the maximum effect on ecosystem quality (5.40 points) and resources (12.50 points) followed by transportation phase, which have 4.51 points and 8.34 points impacts in same categories. The maximum impact in manufacturing phase is due to the resource consumption. However, transportation phase (8.23 points) has the highest impact on human health followed by manufacturing phase (6.34 points). The major impacts in human health during transportation phase may be attributed to the release of toxic gases like carbon monoxide, sulfur dioxide and nitrogen oxides

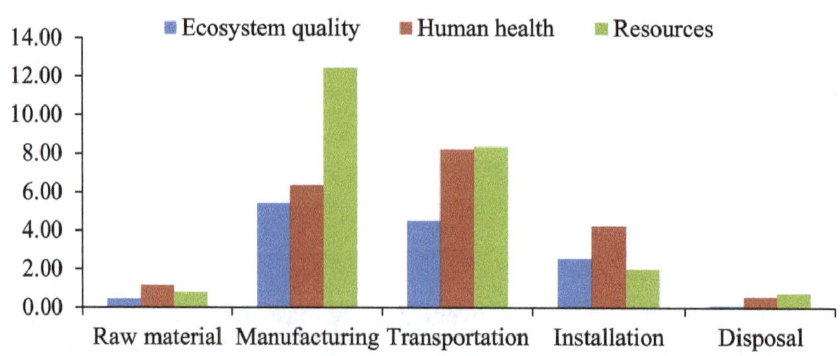

Fig. 17.2 Endpoint environmental impacts in different phases

due to the burning of fuel. The effects of installation phase, though less than manufacturing and transportation phases, are still significant. Installation phase have 2.55 points impacts on ecosystem quality, 4.23 points impacts on human health, and 1.99 point impacts on resource depletion. Raw material extraction phase has very low effect on the endpoint categories (ecosystem quality 0.46 points, human health 1.14 points, and resource depletion 0.78 point) and the disposal phase has negligible effect (ecosystem quality = 0.11 points, human health = 0.58 points, and resource depletion = 0.76 point). This is because the disposal of sanitary ware does not produce any harmful or toxic emissions in any form. Moreover, sanitary ware manufacturer avoids use of redoxide during glazing, which further reduces the environmental impacts in raw material and disposal phases.

Figure 17.3 shows the environmental impact assessment using ReCiPe endpoint assessment method for the used resources during the life cycle of a sanitary ware. It shows the impact for various materials, energy and facilities (feldspar, clay, heavy fuel oil, inert waste, grass, electricity, industrial machine, transport, etc.) on the three endpoint impact categories. It shows that the impact caused by the manufacturing phase on ecosystem quality is mainly due to the use of grass in packaging (3.27 points) followed by electricity consumption (1.35 points) and heavy fuel oil fired in kiln (0.78 points). Impact caused by the manufacturing phase on human health is mainly due to the electricity consumption (3.41 points) followed by heavy fuel oil (1.48 points) and grass (1.46 points). Impact caused by the manufacturing phase on resources is mostly due to the heavy fuel oil usage (9.83 points). The studies conducted by Ye et al. (2018). Özkan et al. (2016), and Sangwan et al. (2018) observed similar results and identified that fuel for firing, electricity, and packaging material have high environmental impacts during the manufacturing phase of ceramic products.

Industrial machines used in the manufacturing phase have insignificant effect on the three categories. Cement mortar causes the impacts during the installation phase and inert waste in the disposal phase. All the materials in the raw material extraction phase (feldspar, ball clay, china clay, quartz, water, zirconium, zinc, soda ash, silicate, gypsum, and barite) have very low environmental impacts in the selected endpoint categories. The highest impact is due to the use of trucks in transportation of sanitary ware. The ceramic industries have clusters in specific region of India based on the

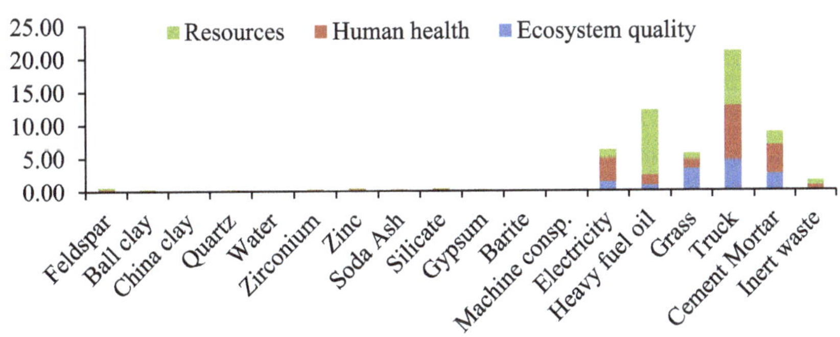

Fig. 17.3 Endpoint environmental impacts by different materials

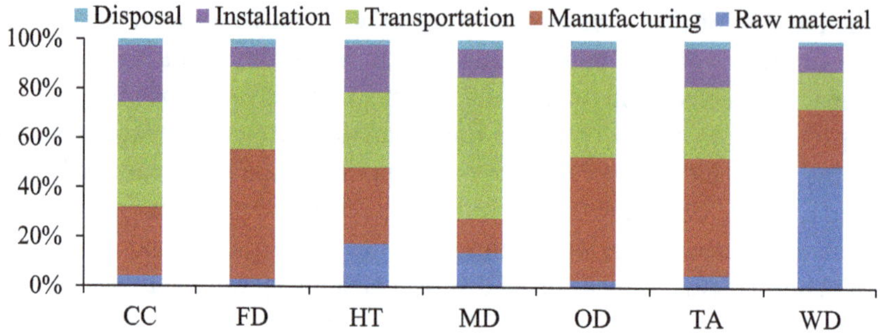

Fig. 17.4 Midpoint environmental impacts in different phases

availability of raw material. The transportation of products from these clusters is through trucks (lorries) which is not a sustainable option. The opening of future industries should be near to the market as the transportation of the raw material has much less environmental impact because of low volume compared to the finished goods.

17.4.2 Midpoint Assessment

Figure 17.4 show the environmental impact assessment using ReCiPe midpoint assessment method. It depicts the percentage-wise impact of each phase in the categories of climate change (CC), fossil depletion (FD), human toxicity (HT), metal depletion (MD), ozone depletion (OD), terrestrial acidification (TA), and water depletion (WD).

The major contributors in all the categories, except water depletion, are the manufacturing (CC = 136 kg CO_2-Eq, FD = 115 kg oil-Eq, HT = 26.9 kg 1,4-DCB-Eq, MD = 2.70 kg Fe-Eq, TA = 1.38 kg SO_2-Eq, and WD = 0.34 m^3) and transportation phases (CC = 209 kg CO_2-Eq, FD = 72.7 kg oil-Eq, HT = 26.7 kg 1,4-DCB-Eq, MD = 11.40 kg Fe-Eq, TA = 0.84 kg SO_2-Eq, and WD = 0.22 m^3). In water depletion category, raw material extraction phase (0.72 m^3) contributes almost 50% environmental impact. Manufacturing phase contributes almost 50% impact in fossil depletion (115 kg oil-Eq) and terrestrial acidification (1.38 kg SO_2-Eq) categories due to the use of electricity (FD = 16.7 kg oil-Eq and TA = 0.45 kg SO_2-Eq) and heavy fuel oil (FD = 91.1 kg oil-Eq and TA = 0.23 kg SO_2-Eq) in kiln firing. Transportation phase contributes almost 56% in metal depletion category (MD = 11.4 kg Fe-Eq). Installation has around 26% impact on the climate change category (113 kg CO_2-Eq) due to use of mortar. Disposal phase has the least impact in all the seven categories (CC = 13.2 kg CO_2-Eq, FD = 6.77 kg oil-Eq, HT = 1.63 kg 1,4-DCB-Eq, MD = 0.70 kg Fe-Eq, TA = 0.08 kg SO_2-Eq, and WD = 0.02 m^3) because sanitary ware is an inert waste, which is not toxic and harmful.

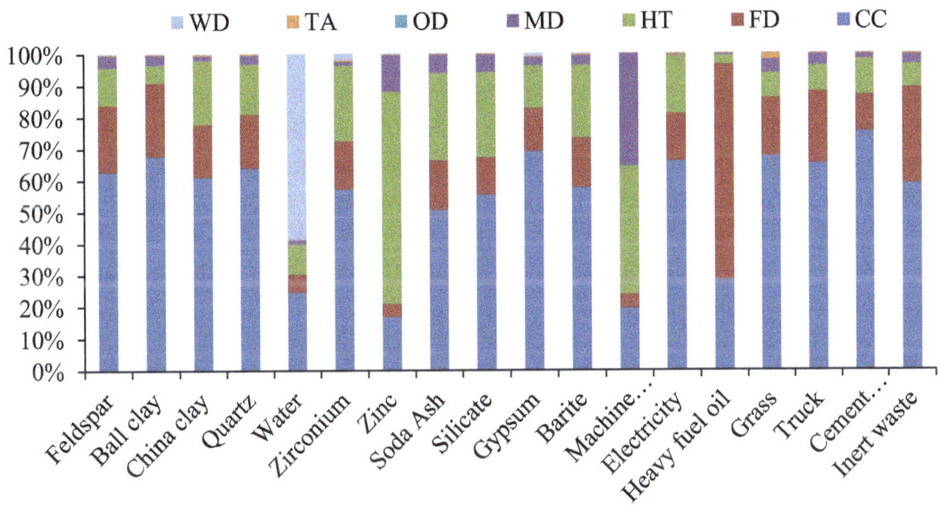

Fig. 17.5 Midpoint environmental impacts by different materials

The results of ReCiPe midpoint assessment method by different materials, energy and facilities are also shown in Fig. 17.5. It shows the percentage-wise contribution of the seven categories from various materials, energy and facilities used along the supply chain. The materials, energy and facilities (except water) have major impacts in climate change category.

Zinc contributes around 67% to human toxicity (9.21 kg 1,4-DCB-Eq), industrial machines contribute around 40% to human toxicity, and heavy fuel oil contributes around 67% to fossil depletion (91.1 kg oil-Eq). Cement mortar used for installation of sanitary ware has the maximum climate change potential (113 kg CO_2-Eq). Industrial machines used in manufacturing phase cause the maximum depletion of metals.

However, Fig. 17.5 shows the effects in relative percentage term, it does not reflect the effect in absolute terms. The effect of industrial machines consumption only shows that the human toxicity, metal depletion and climate change are the main affected categories out of the seven categories. The total effect of machine consumption is almost negligible compared to other resources used.

17.5 Practical Implications

This study identifies that heavy fuel oil, packaging material, electricity consumption, cement mortar, and transport of the finished products are the hotspots of the environmental impacts in manufacturing, installation, and transportation phases. The findings of the study can be used by the policy makers to form the future strategies for this sector to reduce the environmental impacts. The findings of this study will generate awareness among the manufacturers of this sector and help the government to form the regulations for this sector. Awareness and legislation should be focused simultaneously for the reduction of environmental impacts from the construction sector. The future legislations should focus on the environmental friendly use and production of construction materials, establishment of new enterprises nearer to the market. Government should also conduct workshops and seminars to educate and aware the top management of the construction enterprises about the latest technologies, environmentally efficient processing, and efficient resource utilization to decrease the environmental impacts.

This study suggests the replacement of heavy fuel oil by natural gas in kiln firing, renewable energy source of electricity, and alternate material in packaging to reduce impact in the manufacturing phase. This study recommends replacing the cement from the mortar with marble slurry in certain percentage to reduce the environmental impacts during installation phase. There is a possibility to use ceramic waste in the production of concrete, therefore future research should be focused in this direction. This will help to reduce the use of virgin materials and also solve waste disposal problem of ceramic sector. The methodology, assessment methods and impact categories used in the study can be used by other ceramic enterprises for the identification of environmental hotspot in their supply chain. Also, the results of the study can be used by the different ceramic enterprises for benchmarking purpose.

17.6 Conclusions

This study has assessed, compared and analyzed the results of the environmental impacts caused in different phases by the materials and resources used in a sanitary ware supply chain. It is found that manufacturing and transportation phases have high environmental impacts followed by installation phase. The other two phases of this supply chain (raw material extraction and disposal) cause insignificant damage to the environment. It is also found, among the materials used throughout the supply chain, that heavy fuel oil, electricity, grass, and cement mortar cause maximum damage to the environment. Hence, this study will help the policy makers to act on the identified hot spots of environmental impacts in sanitary ware supply chain and take appropriate actions to develop future green strategies for this industry to enhance sustainability. These findings suggest that there is a need to focus on environmental efficiency in the manufacturing and transportation phases of the supply chain. Sanitary ware industries should be distributed around the country to reduce the environmental impact caused by transportation.

The limitation of this study is that the primary data has been collected from only one manufacturing plant in India. The study represents the whole sanitary ware supply chain by considering selected case enterprise as a focal firm. However, this study does not represent the multiplayers at the each stage of the supply chain. Future studies can consider more than one case and use this study for the benchmarking purpose. Also, the sanitary ware waste after useful life is taken as inert waste, which is nearest in properties in ecoinvent 3.0 database. This study does not consider transportation from retailers to customers because of complexity in data collection. Moreover, the distances involved in transportation from retailers to customers are comparatively negligible.

References

Balasbaneh AT, Bin Marsono AK, Kasra Kermanshahi E (2018) Balancing of life cycle carbon and cost appraisal on alternative wall and roof design verification for residential building. Construct Innov 18(3):274–300. https://doi.org/10.1108/CI-03-2017-0024

Bathroom & Kitchen Expo Overview of India's Bathroom & Kitchen Industry. http://www.gima.de/media/1/1/4798/overview_of_india_s_bathroom_kitchen_industry.pdf. Last assessed 13 Apr 2018

Bhamu J, Sangwan KS (2015) Reduction of post-kiln rejections for improving sustainability in ceramic industry: a case study. Procedia CIRP 26:618–623. https://doi.org/10.1016/j.procir.2014.07.176

Bribián IZ, Capilla AV, Usón AA (2011) Life cycle assessment of building materials: comparative analysis of energy and environmental impacts and evaluation of the eco-efficiency improvement potential. Build Environ 46(5):1133–1140. https://doi.org/10.1016/j.buildenv.2010.12.002

Cabeza LF, Rincón L, Vilariño V, Pérez G, Castell A (2014) Life cycle assessment (LCA) and life cycle energy analysis (LCEA) of buildings and the building sector: a review. Renew Sustain Energy Rev 29:394–416. https://doi.org/10.1016/j.rser.2013.08.037

Dong YH, Ng ST (2014) Comparing the midpoint and endpoint approaches based on ReCiPe—a study of commercial buildings in Hong Kong. Int J Life Cycle Assess 19:1409–1423. https://doi.org/10.1007/s11367-014-0743-0

Guinee JB, Heijungs R, Huppes G, Zamagni A, Masoni P, Buonamici R, Ekvall T, Rydberg T (2011) Life cycle assessment: past, present, and future. Environ Sci Technol 45(1):90–96

Mittal VK, Sangwan KS (2014) Development of a structural model of environmentally conscious manufacturing drivers. J Manufact Technol Manage 25(8):1195–1208. https://doi.org/10.1080/00207543.2013.838649

Özkan A, Günkaya Z, Tok G, Karacasulu L, Metesoy M, Banar M, Kara A (2016) Life cycle assessment and life cycle cost analysis of magnesia spinel brick production. Sustainability (Switzerland) 8(7):1–13. https://doi.org/10.3390/su8070662

Rebitzer G, Ekvall T, Frischknecht R, Hunkeler D, Norris G, Rydberg T, Schmidt WP, Suh S, Weidema BP, Pennington DW (2004) Life cycle assessment Part 1: framework, goal and scope definition, inventory analysis, and applications. Environ Int 30(5):701–720. https://doi.org/10.1016/j.envint.2003.11.005

Sangwan KS, Choudhary K, Batra C (2018) Environmental impact assessment of a ceramic tile supply chain—a case study. Int J Sust Eng 11(3):211–216. https://doi.org/10.1080/19397038.2017.1394398

Ye L, Hong J, Ma X, Qi C, Yang D (2018) Life cycle environmental and economic assessment of ceramic tile production: a case study in China. J Clean Prod 189:432–441. https://doi.org/10.1016/j.jclepro.2018.04.112

18

Indian Economy and the Role of the Cement Industry

Kapil Kukreja, Prateek Sharma, Bibekananda Mohapatra, and Ashutosh Saxena

Abstract Indian cement industry is rated as one of the best performing industry across various industrial sectors in terms of energy efficiency, quality control, environmental sustainability and adaptive in venturing into new technology. Indian cement industry is contributing to circular economy primarily by (i) Circular Supply Chain, (ii) Recovery and Recycling. Waste from various industries is being utilized by the cement industry as alternative fuels and raw materials (AFR). As cement manufacturing process itself supports the environmentally sustainable waste utilization due to high temperature incineration without leaving any residue, hence it is acting as backbone for waste generating industries. National Council for Cement and Building Materials (NCB) being a leading R&D organization in the field of Cement and Building Materials in India is working to support the cement industry to enhance the waste utilization and sustainable manufacturing for clean and green India. This paper highlights the efforts of Indian cement industry, contribution of NCB towards circular economy and futuristic potential.

Keywords Recovery and recycling · Renewable energy · Waste utilization

18.1 Introduction

India is among world's fastest growing economy showing resilient to external factors. The economies of scale have predominantly taken over the narrative around resource use leaving principles of circularity and resource efficiency in the background. The long-term growth perspective is high but with the rise in resource demand. The country's natural resources are under strain and there is critical need for resource efficiency improvement. Circular economy is emerging approach, which can take the country to newer heights without straining the resource supply. Circular economy looked towards the elimination of any kind of waste in the market. It defines waste to

K. Kukreja (✉) · P. Sharma · B. Mohapatra · A. Saxena
Centre for Mining, Environment, Plant Engineering and Operation, National Council for Cement and Building Materials, Ballabgarh 121004, HR, India
e-mail: kkancb@gmail.com

any kind of underutilization of resources or assets rather than its interpretation as junk material (https://niti.gov.in/writereaddata/files/E-WasteStrategy.pdf). The challenge to put circular consumption into practice can be addressed by 3R Principle that is based on Reduce, Recycle and Reuse (Ghosh 2017). There are five streams of circular economy models i.e. (i) Circular Supply Chain; Provide renewable energy, bio-based or fully recyclable input material to replace single-lifecycle inputs, (ii) Recovery and Recycling; Recover useful resources/energy out of disposed products or by-products, (iii) Product Life Extension; Extend working lifecycle of products and components by repairing, upgrading and reselling, (iv) Sharing Platform; Enable increased utilization rate of products by making possible shared use/access/ownership, (v) Product as a Service; Offer product access and retain ownership to internalize benefits of circular resource productivity (https://Circular%20economy/FICCI-Circular-Economy.pdf). Cement industry has been considered as one of the pillar of growth for any nation. Indian cement industry being one of the second largest cement production after China with an installed capacity of 509 MTPA (million tonnes per Annum) in 2018 is constantly contributing for the circular economy of India by various means and this paper highlights Indian cement industry gains through circular economy as well as future potential (https://pib.gov.in/Pressreleaseshare).

18.2 Achievements of Indian Cement Industry in Last Decade

Indian cement industry is rated as one of the best performing industries across various industrial sectors in terms of energy consumption, quality control, environmental sustainability and adaptive in venturing into new technological options. Some of the recent major strides of Indian cement industry are reduction of CO_2 emission factor from 1.12 t of CO_2/t of cement in 1996 to 0.670 t of CO_2/t of cement in 2017, enhanced blended cement production from 68% in 2010 to 73% of total cement production in 2017. Thermal Substitution Rate (TSR) is part replacement of conventional fuel by alternative fuels in terms of thermal energy requirement and is calculated as percentage of heat supplied by alternative fuel from the total heat requirement for pyro-processing in a cement plant. % TSR has improved to 4% now as compared to a dismal 1% only 3–4 years back. Cement plants have adopted technologies to meet the new emission norms for Particulate Matter (PM) and NO_X emissions. Plants have installed high efficient bag filters, Electrostatic precipitators (ESPs), hybrid filters to control dust emissions. For NO_X reduction, plants have installed secondary control measures like Selective Non Catalytic Reduction (SNCR). All the cement plants have installed continuous emission monitoring system as per the guidelines of Central pollution Control Board. Indian cement sector is most energy efficient worldwide, mainly due to modern technology being implemented in the plants as well as because of efficient monitoring of plant's performance on a daily basis, focusing on energy savings and CO_2 emissions reduction. Indian cement industry growth in next decade looks very promising. Cement demand is projected to grow to 2.5 to 2.7 times the current volumes and reach 550 to 600 MTPA by 2025 (https://www.wbcsd.org/Sector-Projects/Cement-Sustainability-Initiative/Resources/Low-Carbon-Technology-Roadmap-for-the-Indian-Cement-Sector-Status-Review). Per capita consumption is likely to increase from 210 to 580 kg world average. The government of India has launched various new urban development missions including development of 500 cities, setting up of 100 Smart Cities in the country by 2022, Affordable housing under "Housing for All till 2022" and dedicated freight corridors etc. (https://www.ibef.org/industry/cement-presentation).

18.3 Contribution of Indian Cement Industry in Circular Economy

Cement industry contribution to circular economy is primarily under two heads i.e. (i) Circular Supply Chain, (ii) Recovery and Recycling. Figure 18.1 indicates that how the cement industry is contributing in circular economy and sustainable manufacturing.

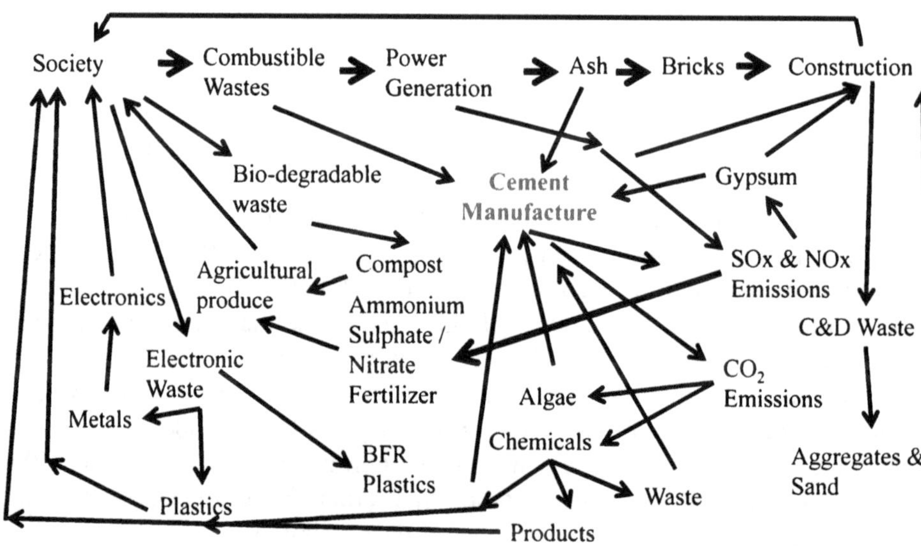

Fig. 18.1 Role of cement manufacturing in circular economy

18.3.1 Circular Supply Chain

In this head of circular economy, Indian cement industry is playing a key role by enhancing the application of Renewable Energy for electrical power generation. The renewable energy installed capacity (wind and solar) in cement plants increased by more than 40% to 276 MW from 2010 to 2017. Out of the total, 42 MW is solar power, while off-site wind installations account for 234 MW. A company has undertaken the target of switching over to renewable energy for 100% of all electrical energy needs by 2030 (https://www.wbcsd.org/Sector-Projects/Cement-Sustainability-Initiative/Resources/Low-Carbon-Technology-Roadmap-for-the-Indian-Cement-Sector-Status-Review). Big players like UltraTech Cement are targeting 25% share of their total power consumption by green energy technologies by 2021 (https://www.thehindubusinessline).

Apart from solar photovoltaic route, cement industry is making efforts to tap solar energy through thermal route. A study has been undertaken in Europe for solar reactor design operating at 800–1000 °C, using rotary kiln and a horizontal bubbling fluidized bed, to manufacture cement (Moumina et al. 2019; https://www.solarpaces.org/new-iea-report/). Another study presents the design of a mini, scalable solar lime kiln which was designed using solar dish collector to calcine small sized (1–5 mm diameter) limestone particles. The heat is focused on a heating element located centrally in a tilted rotary kiln driven by chain drive (Swaminathan and Nadhipite 2017). Some studies have been done for feasibility study for Concentrated Solar Thermal technology in cement industry (Gonzalez and Flamant 2014).

18.3.2 Recovery and Recycling

It has been established that different types of wastes/by products of other industries available worldwide can be utilized as alternative fuels and raw materials for cement production. Moreover, production of blended cements, composite cements and utilizing performance improvers in cement also support circular economy. Use of fly ash and granulated blast furnace slag (GBFS) in the production of blended cements i.e. Portland Pozzolana Cement (PPC) and Portland Slag Cement (PSC) is also beneficial for conservation of natural resources, lowering in clinker factor in cement and reduction in CO_2 emissions along with environmental sustainability. Contribution of Indian Cement Industry in Circular Economy along with their associated challenges are highlighted in following Table 18.1.

Table 18.1 Contribution in circular economy and associated challenges

Contribution in circular economy	Associated challenges
Fly ash based cement: Flyash is the solid waste generated in thermal power plants. Out of the total fly ash generation, around 25% is being utilized for cement industry (http://www.cea.nic.in/reports/others/thermal/tcd/flyash_201617.pdf)	33% around still remains unutilized due to Geographical imbalanced and limitation of maximum 35% fly ash, in PPC, as per IS:1489 (Part-I) (http://www.cea.nic.in/reports/others/thermal/tcd/flyash_201617.pdf)
Alternative fuels: The cement manufacturers are consuming all possible Alternative Fuels (AFs) like refuse-derived fuel (RDF), industrial plastic, biomass, tyre chips, waste generated by Pharmaceutical industry, Paint industry, Agro industry, Paper industry, chemical industry etc. (Mohapatra et al. 2014; Shaw et al. 2017)	Waste management, Economic viability, Geographically Availability, Heterogeneous nature of waste, Capital investment, no existence of polluter pay principal (Julie 2004)
Slag based cement: Blast Furnace Slag (BFS) and Steel Slag is the solid waste generated in Iron and Steel industry. Currently India produced approx. 25 million tonnes BFS out of which 22 million tonnes of BFS is granulated and being consumed entirely in cement industry (Agarwal et al. 2017)	Steel slag is still remains unutilized (Agarwal et al. 2017)
Waste heat recovery (WHR): Waste heat obtained from flue gases and air of preheater and clinker cooler respectively is utilized to generate electricity. Indian cement industry has implemented WHR systems to produce power and WHR capacity has gone up 212% to 344 MW from 2010 to 2017 (https://www.wbcsd.org/Sector-Projects/Cement-Sustainability-Initiative/Resources/Low-Carbon-Technology-Roadmap-for-the-Indian-Cement-Sector-Status-Review)	High capital costs of the system, need to be offset with appropriate government policies including renewable status for WHR systems, PAT incentives etc. (https://pib.gov.in/Pressreleaseshare)

18.4 NCB's Experiences

NCB being a premier R&D organization for cement and construction sector in Inida has executed a number of R&D and consultancy projects related to waste utilization and energy conservation. Outcomes of these project is helping to Indian cement industry for contributing towards the circular economy and sustainable manufacturing. Some of them are highlighted below.

18.4.1 Production of Synthetic Slag from Low Grade Limestone

A Study were carried out at NCB laboratory for development of Synthetic Slag using low-grade limestone. Laboratory slag samples prepared with low-grade limestones and other additive materials, which found to be conforming the IS: 12089-1987. These laboratories made synthetic slag samples as shown in Fig. 18.2 were also investigated by optical microscopy as shown in Fig. 18.3 and found to have maximum 92% glass content, which is greater than 85% as specified in IS-12089. PSC samples were prepared with 40 and 60% synthetic slag replacing equal quantity of clinker. The performance of PSC blends prepared using synthetic slag sample were found as per requirements of Indian Standard Specification, IS: 455-1989 for PSC. As the limestone, which is getting depleted and has reached to an alarming level where the availability of cement grade limestone in India has reduced to 8949 million tones only, Synthetic Slag may play a vital role to replace clinker or indirectly cement grade limestone. However, main challenge would be to produce this slag at industrial scale.

Fig. 18.2 Synthetic slag preparation in NCB laboratory

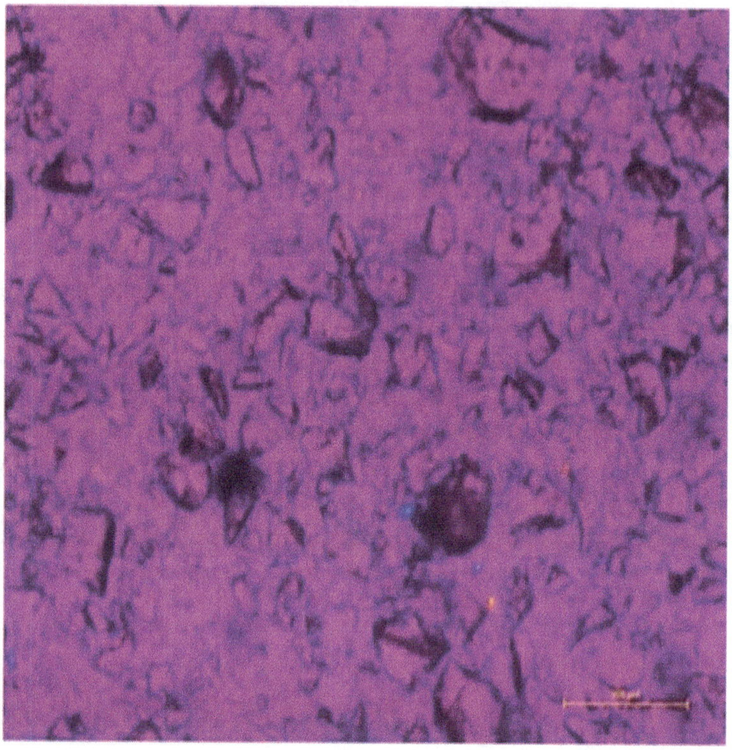

Fig. 18.3 OM analysis of water quenched laboratory synthetic slag sample, glass content-92%

18.4.2 Alternative Fuels Utilization

NCB has vast experience of analytical studies, trial runs and system design for Alternative Fuels (AFs) utilization. NCB has carried out various studies, which covers the impact assessment of AFs on overall process. Recently, number of studies have been carried out for cement plants. One study for utilization of Tyre Derived Fuel (TDF) as alternative fuel for co-processing. TDF ash sample was tested at NCB laboratory, and Zinc content was found to be around 4.06%, which is equivalent to 0.03% by weight in clinker. Several investigations have already revealed that zinc concentration above 500 ppm in clinker impact the workability of cement. Considering the impact of addition of Zinc in clinker, plant can achieve around 21% Thermal Substitution Rate (TSR) however, TSR beyond 21% may not be feasible due to higher zinc content in TDF. Another study was done for a cement plant (located in North Karnataka, India) to handle more than 25 types of alternative fuels, plant was commissioned a year back and now utilizing waste as a fuel with more than 10% TSR. NCB is also doing projects for system design to handle five different types of alternative fuels for a cement plant located in South & Central part of India.

18.4.3 Dolomitic Limestone Utilization

NCB in one of its recent projects has successfully utilized of up to 15% dolomite as an additive replacing equal quantity of clinker. The cement performance was found to bc similar to that of control cement prepared without dolomite.

18.4.4 Performance Improvers in Cement Manufacturing

Based on studies on number of industrial wastes by NCB, the Bureau of Indian Standards (BIS) has permitted the use of copper slag, LD slag, lead-zinc slag and catalytic waste from petroleum refinery as performance improvers in manufacture of OPC.

18.4.5 Bottom Ash Utilization

In a recent R & D work, NCB has developed tiles and bricks by utilization of ~30% bottom ash along with fly ash. Cementitious binders with consistent strength property were prepared using rationalized formulations and curing conditions. The pre-cast bodies like tiles (150 × 75 × 25 mm) were meeting test limits of IS 2690 (Part 2):1992 and bricks having dimensions of 190 × 90 × 40 mm were meeting the requirement of strength and water absorption of Class 15 of IS 1077 (Part 2):1992. In another research activity at NCB, Bottom Ash is also successfully utilized as 50% replacement of fine aggregate in concrete.

18.4.6 Composite Cement Production

NCB has carried out several studies on composite cement wherein combinations of fly ash and granulated blast furnace slag were used for preparing composite cement blends. BIS has brought out standard specification IS: 16415-2015 for composite cement on recommendations of NCB.

18.4.7 Alternative Raw Materials Utilization

NCB has done an investigative study to utilize 15 inorganic industrial wastes including limes sludge, Wolstanite, leather sludge, Jarosite, LD slag, red mud and marble slurry in cement manufacture and as aggregate in concrete.

Table 18.2 Country wise alternate fuel utilization (million tons) (http://www.ciiwasteexchange)

S. No.	Country name	Value
1.	Germany	3.1
2.	North America	2.1
3.	USA	1.9
4.	India	1.6
5.	Brazil	1.3

18.5 Comparison with Other Countries

One of the major aspects of circular economy for Indian cement industry is Alternative Fuels & Raw Materials (AFR). India's Thermal Substitution rate is comparable with other countries such as 100% TSR in Australia and 100% TSR in France. Following Table 18.2 indicates that India has fourth position in the world to utilize Waste as Fuel on the basis of Quantity.

Above table indicates that waste utilization as fuel in India is still reasonable good considering high cement production capacity but still has huge potential to achieve high TSR. In terms of WHR, Cogeneration systems are well established in cement industry all over the world with Japan, China, India and Southeast Asian countries taking the lead in this development (http://www.ciiwasteexchange).

18.6 Futuristic Scenario

Development of Portland Composite Cement (Fly ash/Slag and Limestone based), Development of Portland Limestone Cement (PLC), utilization of low grade limestone and mines rejects, Utilization of Construction and demolished waste (C&D) waste based aggregates in concrete structures and pavements are some of the key areas, where Indian Cement Industry & NCB is working together towards resource conservation and boost to circular economy in India (Table 18.3).

Table 18.3 Future potential for AF utilization by Indian cement industry (http://www.ciiwasteexchange.org/doc/annexure_6.pdf)

Waste streams	% Share on AF	% Share on thermal energy
MSW	57.07	14.27
Spent pot lining	0.81	0.20
Biomass	33.97	8.49
Hazardous waste	3.46	0.87
Tyre waste	7.33	1.83
Total		25.66

Table 18.4 Annual thermal heat utilization by Indian cement industry (https://pib.gov.in/Pressreleaseshare)

Parameter	Value	Unit
Cement production	297.50	MTPA
Clinker factor	0.71	
Clinker production	211.22	MTPA
Fuel consumption (including coal and petcoke)	39.00	MTPA
Fuel consumption (in terms of % clinker)	18.43	%
Average calorific value of fuel	5900	kcal/kg fuel
Total heat utilized	23×10^7	Million kcal/annum

As a latest development, Indian government is planning to ban the single use plastic very soon. Government is looking towards Indian cement industry to burn the existing plastic waste and the industry is quite capable to do so due to some of the inherent features of cement manufacturing process. A typical analysis of entire single use plastic waste consumption in Indian cement industry as fuel considering existing production and fuel usage is shown in Table 18.4.

It is encouraging to see in Table 18.5 that % TSR of the Indian cement industry can go up by 5.5% with overall TSR of around 9.5% by utilizing 90% of single use plastic as fuel and replacing conventional fuel like coal and petcoke. This will provide a steady path to achieve 25% TSR by year 2025 and will encourage the circular economy in near future (http://www.ciiwasteexchange.org/doc/annexure_6.pdf).

Another potential area for Indian Cement industry is Geopolymer cements; Geopolymeric cements are eco-friendly binders and being produced from non-limestone bearing raw materials and wastes such as fly ash and slag. Thermal Power plants (TPP) in India are also in the process of installation of Flue Gas Desulphurization (FGD) systems to control SOx emissions. By product of this system is FGD gypsum which can be a partial/fully replacement of natural gypsum used to control setting

Table 18.5 Anticipated % TSR by utilizing single use plastic by Indian cement industry

Parameter	Value	Unit
Total plastic waste	9.40 (Raju 2019)	MTPA
Non-recyclable	3.76 (Raju 2019)	MTPA
Available for co-processing	3.40	MTPA
Average calorific value of fuel	3750	kcal/kg plastic
Total heat available	1.27×10^7	Million kcal/annum
Potential thermal substitution rate	5.50	%

time of Portland cements (Caillahua and Moura 2018). Cements plants in India, which are facing the issue of gypsum availability, may procure FGD gypsum from power plants in future.

18.7 Conclusions

Indian cement industry has to play a catalyst role in future towards resource conservation and providing impetus to circular economy. All stakeholders including cement plants, research organizations (Like-NCB), Government Bodies (Pollution Control Boards, Municipal Corporation etc.) etc. have to work together in one direction to achieve goals of circular economy. In coming years, Circular economy will gain further momentum in Indian cement industry by utilizing Gypsum generated from FGD in TPP, Consumption of Non-recyclable plastic waste, Production of High Volume Flyash cement, Utilizing Steel slag, reduction clinker factor by Alternative raw materials and increasing TSR by the use of AFs. However, Solar Thermal Calcination and Geopolymer cement will take time to establish. Some newer avenues need to be explored in this area like product as a service where cement industry can also work out to buy compressed air and other utilities instead of procuring compressors, pumps etc. It may lead to opportunities of futuristic technologies like oxy fuel combustion. Some companies can sell pure oxygen to cement plants cluster at a reasonable price. Another interesting aspect of circular economy, which remains unexplored in context of cement industry is product life extension, at present uses of C&D waste is taking momentum which helps form product life cycle extension.

References

Agarwal SK, Vanguri S, Chaturvedi SK, Kumar A, Reddy AS (2017) Performance evaluation of granulated BF slag—steel slag based Portland slag cement. In: 15th International seminar on cement, concrete and building materials, New Delhi

Caillahua MC, Moura FJ (2018) Technical feasibility for use of FGD gypsum as an additive setting time retarder for Portland Cement. J Mater Res Technol 7(2):190–197

CII Homepage. http://www.ciiwasteexchange.org/doc/afr2018.pdf. Last accessed 8 June 2019

CII Homepage. http://www.ciiwasteexchange.org/doc/annexure_6.pdf. Last accessed 8 June 2019

CEA Homepage. http://www.cea.nic.in/reports/others/thermal/tcd/flyash_201617.pdf

FICCI Homepage. http://ficci.in/spdocument/22977/FICCI-Circular-Economy.pdf. Last accessed 12 June 2019.

Ghosh SK (2017) Circular economy and 3Rs—reduce, reuse, recycle, for efficient use of resources in the Asia and the Pacific. In: Proceedings of the WasteSafe 2017—5th international conference on solid waste management in the developing countries, 25–27 February 2017, Khulna, Bangladesh

Gonzalez RS, Flamant G (2014) Technical and economic feasibility analysis of using concentrated solar thermal technology in the cement production process: hybrid approach—a case study. J Solar Energy Eng 136:025001-1 to 12

IBEF Homepage. https://www.ibef.org/industry/cement-presentation. Last accessed 18 Oct 2019

Julie D (2004) Resource recovery, SSW synthesis part 3, good practices and recommendations on SSW use. Julie, Lafarge Res Recov 26–28

Mohapatra BN, Vyas SK, Shekhar C (2014) Indian experience of using AFR. In: Cement kiln. Industrial Angles, vol 3, pp 7–13

Moumina G, Tescarib S, Sundarrajb P, Oliveirab L, Roebb M, Sattlerb C (2019) Solar treatment of cohesive particles in a directly irradiated rotary kiln. Solar Energy 182:480–490

NITI Aayog Homepage. https://niti.gov.in/writereaddata/files/E-WasteStratcgy.pdf. Last accessed 15 June 2019

PIB Homepage. https://pib.gov.in/Pressreleaseshare.aspx?PRID=1562754. Last accessed 15 June 2019

Raju A (2019) Plastic waste management and marine plastic pollution in South Asia. In: Second sub-regional workshop on preparation of status report and sub-regional roadmap for implementing the global waste management goals towards addressing SDGS in South Asia, Dhaka

SolarPACES Homepage. https://www.solarpaces.org/new-iea-report/. Last accessed 8 June 2019

Shaw SK, Saxena A, Kukreja K, Rao MVR, Venkatesh V (2017) Feasibility study for enhancement of solid alternate fuels in Indian cement plant. In: Proceeding of the 15th NCB international seminar on cement and building materials, New Delhi

Swaminathan R, Nadhipite S (2017) Design of solar lime kiln. Innov Energy Res. https://doi.org/10.4172/2576-1463.s1-001

The HINDU Business Line Homepage. https://www.thehindubusinessline.com/companies/by-2021-ultratech-aims-to-source-25-of-energy-from-renewable-sources/article27689331.ece. Last accessed 12 June 2019

WBCSD HomePage. https://www.wbcsd.org/Sector-Projects/Cement-Sustainability-Initiative/Resources/Low-Carbon-Technology-Roadmap-for-the-Indian-Cement-Sector-Status-Review-2018. Last accessed 12 June 2019

19

Design Thinking based Teaching: Enhancing Engineering Education

Sangeeta Sharma and Priya Christina Sande

Abstract Teaching is an art which requires systematic exploration in manifestations, be it curriculum design, pedagogical strategies and theories, teaching material and practices, and technological interventions. Due to the interface of technology and the mitigating attention span of the students, teachers have realized that the old method of teaching is not relevant anymore. As a result in the different courses offered at the Engineering institute, the teachers are adopting divergent thinking, synectics and design thinking for better understanding of the concepts. Design thinking offers better solutions to the problems and includes analogies and synectics to clarify the concepts. Design thinking aids better student participation and fosters teacher-student relationship. The students are encouraged for team based learning, which make them more attentive. This paper discusses how design thinking has been applied in two popular courses at Birla Institute of Technology and Sciences, BITS Pilani namely General Biology and Computational Fluid Dynamics. The feedback suggests that through analogies and design thinking the profound concepts are made easy to grasp.

Keywords Engineering education · Design thinking · Innovative pedagogy

19.1 Introduction

Teaching can be considered a science as well as an art which requires systematic research, exploration and innovation along with constant development and adaptation in methods and practices. Even with a static curriculum, the strategies of teaching can be labile with the purpose of constantly engaging (Coe et al. 2014). To be effective,

S. Sharma (✉)
Department of Humanities and Social Sciences, Birla Institute of Technology and Science, Pilani-Campus, Pilani 333031, RJ, India
e-mail: sang@pilani.bits-pilani.ac.in

P. C. Sande
Department of Chemical Engineering, Birla Institute of Technology and Science, Pilani-Campus, Pilani 333031, RJ, India
e-mail: priya@pilani.bits-pilani.ac.in

instructors may even integrate multiple strategies over a single course. Knowledge and skills transfer is closely linked with high classroom engagement (Thompson and Irvine 2011). This is as true for engineering courses as it is for any other course.

Even a cursory literature survey shows that research on teaching enhancement in engineering courses have largely revolved around vertical thinking ideas (Vidal 2003). De Bono (1970), who is known as the lateral thinking pioneer of the century, states that "the emphasis has traditionally always been on vertical thinking which is effective but incomplete". He further emphasizes that although analytical thinking is excellent, there is still value addition possible by creative or lateral thinking, which is generative rather than selective in process, and where labels are less water tight than in vertical thinking (De Bono 1970). Educators are constantly grappling with the thought of engaging students fruitfully in learning in the classroom. Design thinking can be employed to solve complex problems faced in all walks of life and different business sectors including education. Herbert (1996) propounded that greater emphasis on cognitive psychology and science of design could help solve complex societal problems.

Design thinking explores new ways to transform the approach to solve such complex problems. It focuses on the needs of the beneficiary, identifies the existing problem in totality and redesigns the models, processes or practices to arrive at a solution. It is fundamentally a user focused process of solving problems innovatively. This process comprises of various elements like observation, collaboration, exploration of latest needs, re-defining of problem statements, visualisation of ideas and concepts, prototyping and organisational change. This work demonstrated the practice of design thinking in the pedagogy of two courses offered at Birla Institute of technology and Science, Pilani. The examples are illustrated from General Biology, a course offered to first year students and Computational Fluid Dynamics, which is offered as an elective course to engineering students.

19.2 Role of Design Thinking

When the students are banking on rote learning, in aspiration to get grades, design thinking carries a lot of importance. It is not only relevant for Science, Technology, Engineering and Maths (STEM) courses but also for humanities as it enables students to think critically and go beyond just asking questions. A few examples could be explanation through synectics, metaphors, divergent thinking and lateral thinking. Edward de Bono had underscored the importance of lateral thinking and interlacing it with humour can lead students think different from familiar patterns to create new patterns. Design thinking has a few well defined elements which have both convergence and divergence in all phases that differentiate between a problem and a solution stage.

Thoring and Muller (2011) defined design thinking process in different stages like understand, observe, define, ideate, select ideas, prototype, test and iterate. Runco and Acar (2012) showed that divergent thinking in the design thinking leads to increased creativity. Divergence in design thinking merely does not emphasize the newly generated idea but also reforming the existing information and knowledge related to the problem (Brown 2009). Divergence relates to the ability to identify multiple solution options to a problem and this then leads to a changed point of view. Convergence utilizes various elements and outcomes of divergent thinking by bringing them together meaningfully through use of methods, patterns, concepts and frame works (Linderberg et al. 2010).

19.3 Design Thinking Application

There are numerous methods used in design thinking. Some of the common characteristics of these methods are: human centricity, collaboration and teamwork, interdisciplinary teamwork, ideation and experimentation (Thoring and Muller 2011). Irrespective of the method used by individual design thinker in solving a problem of any nature, these characteristics are used in one or the other form (Brown 2009). Similarly, design thinkers use different approaches and processes; however, they too manifest certain common attributes while solving problems. They are empathetic, observant and curious, knowledgeable, holistic and integrative thinking, tolerant (deferring judgement), prognostic, experimentalistic and optimistic. The integration of design thinking in the courses emphasizes on team based learning and exchange of ideas amongst students from different engineering field. Unlike traditional method where teacher instructed and students listened, this method encourages people-oriented problem-solving method. Keeping in mind the positive implication of design thinking, where students are given enough liberty to participate in discussions, putting forth their ideas and encouraged for learning through analogies. By doing this the interaction between teacher and students increases and attention also bolsters because of the active participation of the students.

19.3.1 Design Thinking in Biology

Teaching Biology to Engineering students is a challenging task as these students have limited background knowledge of this subject. General Biology is a foundation course offered to all the students at Birla Institute of Technology and Science. Keeping in mind 'creative play', the following were incorporated into the teaching modules which followed standard textbook (Enger and Ross 2011). These examples show how a few basic concepts from a General Biology class made teaching and learning interesting by employing creative activities.

Example 1: The Concept of Neuron Functioning. The concept of neuron functioning is taught by taking example of how the neurons are fixed using the balance of Na+ ions inside and outside of the cells. Na+ concentration is maintained higher outside the cells to maintain resting potential. A group of select students are claimed to be lovely "Sodium Guys" who need to be out of the classroom (classroom represents a neuron) because in resting potential, Na+ ions are outside the cell. After asking the "Sodium Guys" to go out of the class, followed by closing of the door, the door is tapped (which represents physical or chemical signal to the neurons to excite it) and then one part of the door is opened (the door represents the Na+ gated channel) and a few Sodium guys are asked to come inside, once they are in the other part of the door is opened and other "Sodium guys" are asked to get inside. This represent that opening of some of the Na+ gated channels due to physical or chemical induction, allow some Sodium to come inside which in turn allow more Na+ to come inside and thus generating the action potential.

Example 2: The Human Digestive System. When teaching about the Human Digestive System, the reference to "secretion of saliva from salivary gland" is made. As saliva secretion takes place while eating food or just thinking/looking/smelling the food. To explain this, a few chocolate candies' images and other "yummy food" images are shown. The chocolates are distributed to a few students and they are asked to eat. The students who did not take their breakfast were asked to raise their hands. They are then shown images of delicious food. Now they are asked (both who had chocolates and who did not have breakfast) if their mouth is watering. The concept of "Simple Reflex" and "Conditioned Reflex" mechanism of salivary secretion is told. In simple reflex, because the tongue pressure sensor sense the presence of food, it simply passes the signal to "salivary centre" in medulla asking it to send signal to salivary gland for saliva secretion which do not need the help of "Central Nervous System" (CNS). While in case of conditioned reflex (when one is not eating food still have mouth watering) is now visual/olfactory/hearing inputs (looking/smelling/listening about food been processed through CNS to ask salivary control centre in medulla to initiate saliva secretion from salivary glands.

Example 3: The Central Nervous System. To teach about the coordination of hand and leg movement to achieve quicker task, the nodal points present in muscles especially elbow and knee are referred that locally control hand and leg movement. To relate it to the students' passion of cricket, it is asked which students play the game well. Once some students claim that they do, a chalk is thrown at one of them, who then catches the chalk with ease. It is explained on this account that the nodal points of this individual have been trained so well that they can regulate the movement of their hand without the direct involvement of CNS especially for quicker task like catching a ball.

Thus in all the above three concepts of teaching, it has been observed that the use of 'creative activity' increases involvement of the students and the concepts are learnt with ease.

19.3.2 Analogy in Computational Fluid Dynamics: Shock Capturing and Shock Fitting Approaches

As part of the application of design thinking, analogies or more broadly metaphoric activities can help students grasp complex ideas. This is because the analogy creates a mental foothold that can be extended and evolved. In the following example a topic from the course 'Computational Fluid Dynamics' (CFD), is elucidated using an analogy from basic mathematics. This helps define the main CFD idea of shock capturing and shock fitting (Anderson 2012), in a simpler, novel and more interesting manner, rather than directly stating textbook content.

For fast moving objects such as an aircrafts, the fluid flow around the object is affected by shock waves. For example, a bow shock occurs around a blunt nose when the free stream flow is supersonic, or faster than the speed of sound waves (Fig. 19.1). The resulting 'shock' is a front or surface which creates sudden or abrupt change in flow properties of the fluid. The generation of shock waves are due to the fact that fluid molecules ricochet off the moving surface, and stay ahead of the surface, thereby producing pressure or sound waves. When nose approaches sonic velocity there is a piling up or compression of theses sound waves in front of the nose creating the shock.

For aerospace technology, accurate simulation of the surrounding fluid flow containing the shock waves is clearly important. Shock capturing and shock fitting are two CFD approaches to this end. The difference between the two approaches is that in shock fitting, the property values before and after the shock need to be known (from exact relations) and so does the position of the shock. Then CFD only resolves the flow in the region between the shock and body. In shock capturing, CFD resolves

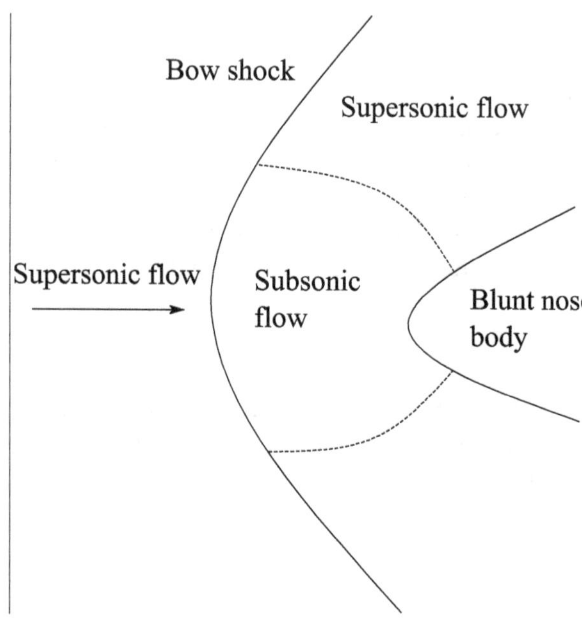

Fig. 19.1 Bow shock around a blunt nose in supersonic flow

the entire flow including the shock wave. To grasp the difference in the philosophy between these two approaches a mathematical analogy is proposed. To follow this analogy the student only needs to know high-school level mathematics. Consider a simple mathematical equation in x and y for which it is required to generate the x, y plot:

$$e^{x} = y * 0.1 \tag{19.1}$$

Consider the following two approaches to accomplish this:

Approach 1. Since the value of constant e is known (2.71828), then for x in any required range, say: 1, 2, 3 … n, the corresponding y is simply calculated as:

$$y = e^{x}/0.1 \tag{19.2}$$

Approach 2. Equation 19.1 is converted (using logarithmic rules) into the form of a straight line like this:

$$\ln y = x - \ln 0.1 \tag{19.3}$$

So on a semi-log graph with ln y on y axis (x as usual on x axis), from the known y-intercept (−ln 0.1 or 2.3), and slope 1, a straight line is plotted. For any x the corresponding y can be extrapolated from the straight line.

Approach 1 is similar to the shock capturing approach in the sense that no other information or theory is required to generate a solution. If Eq. 19.1 is considered analogous to the CFD governing equations (GEs), the direct solution approach used to solve Eq. 19.1 is analogous to shock capturing. The extents of ordinate falls out directly as a result of the calculations, and it is not required a priory to plotting the (x, y) data. In the same way, there is no need to know beforehand the position of the shock wave when using shock capturing approach, because the shock wave is part of the generated solution.

Approach 2 is similar to the shock fitting approach. Here additional theory in the form of logarithmic rules and the form of the straight line are applied. This is analogous to 'extra information' required to calculate flow variables along the explicitly introduced shock wave around the body (Fig. 19.1). In the same way that unknown y values from Eq. 19.1 can be found using the straight line, the introduced shock wave also focuses the CFD solution on the area between body and shockwave. In the same way that slope and intercept are known before y data is generated from Eq. 19.1, the area where the governing equations are required to be resolved is known before the governing equations are actually applied.

19.4 Conclusions

Design thinking is both a tool and a strategic approach. As a tool it provides customer-centric solutions to the products or services through multi-disciplinary approach. Whereas, in strategic approach it helps overcome the magic view and fixed mind set to the problem solution approach and help bringing new energy and thinking to problem solution. It develops collaborative skills, an attitude of questioning, challenging and introspecting within the organisation and individuals gain greater confidence, sharpen their problem solving skills and questions status quo.

For successful integration of design thinking in teaching, it requires shift in thinking orientation in teaching fraternity and the stakeholders must recognize the need for investment in this process to assure long term and sustained success of the Academic Institutes. The examples narrated from two courses namely General Biology and Computational Fluid Dynamics imply that teachers do not leave any opportunity to explain the concepts in most creative way. By doing so, students find themselves responsible to learn and become more active as they are part of the concept as in the case of learning neuron function in biology class.

References

Anderson JD (2012) Computational fluid dynamics: the basics with applications, Indian edn. McGraw Hill, New York

Brown T (2009) Change by Design: how design thinking transforms organization and inspires innovation. Harper Collins, New York

Coe R, Aloisi C, Higgins S, Major LE (2014) What makes great teaching? Review of the underpinning research. Project Report. Sutton Trust, London

De Bono E (1970) Lateral thinking: A textbook of creativity. Ward Lock Educational, London

Enger ED, Ross FC (2011) Concepts in biology, 13th edn. Tata McGraw Hill, New York

Herbert S (1996) The sciences of the artificial, 3rd edn. MIT Press, London

Linderberg T, Gumienny R, Jobst B, Meinel C (2010) Is there a need for a design thinking process? In: Proceedings of the 8th design thinking Research Symposium University of Technology. Sydney, pp 243–254

Runco MA, Acar S (2012) Divergent thinking as an indicator of creative potential. Creativity Res J 24(1), 6–75. https://doi.org/10.1080/10400419.2012.652929

Thompson M, Irvine C (2011) Active learning with the CyberCIEGE video game. In: Proceedings of the 4th conference on cyber security experimentation and test. Scite, San Francisco, p 10. https://doi.org/10.21236/ada547670

Thoring K, Muller RM (2011) Understanding the creative mechanism of design thinking: an evolutionary approach. In: Proceedings of the second conference on creativity and innovation in design. Netherlands, pp 137–147. https://doi.org/10.1145/2079216.2079236

Vidal RVV (2003) Creativity for engineers. Lyngby, Denmark

20

A Study of Metacognitive Awareness in Schindler's List

Punita Raj and Devika

Abstract A grasp over the theoretical concepts directs learners to the comprehension of basic learning but mere acquisition of knowledge does not ensure the path to its applicability in the given situation. Formal education provides engineering students with an environment that allows them to have proper facilities, experts to guide, and well thought -of and critically analysed problems. Awareness of, exposure to, and practice in self-reflective and self-directed learning also known as metacognitive awareness may be put to an incredible use. Engineering education, along with technical aspects, focuses on metacognitive awareness. This qualitative case study attempts to explore knowledge and skills of the character Oskar Schindler in the highly acclaimed real life incident based movie, Schindler's List. Oskar Schindler uses the concept of 'simple engineering' and makes his workers understand or execute a task by helping them understand and implement this concept successfully. This paper also documents the findings of an initiative where Oskar Schindler practically uses his competencies and metacognitive awareness to handle ill-defined or unthought-of problems efficiently, using ignorant and untrained hands.

Keywords Metacognitive awareness · Simple engineering · Competencies

20.1 Introduction

Engineering academic curriculum is mainly influenced by accreditation criteria programs (EA 2005; ABET 2008; ENAEE 2008) that are flexible, developmental and targeted at the specific needs of industries and engineers' expectations (Earnest 2005).

P. Raj · Devika (✉)
Department of Humanities and Social Sciences, Birla Institute of Technology and Science, Pilani-Campus, Pilani 333031, RJ, India
e-mail: devika@pilani.bits-pilani.ac.in

P. Raj
e-mail: p20180428@pilani.bits-pilani.ac.in

In today's professional world, the employees would need metacognitive and self-controlled manner to take situation specific decisions (Conley 2014) and when and how to apply these particular strategies and skills, in doing tasks or solving problems at workplace efficiently (Flavell 1979; Prytula 2012; Sengul and Yasemin 2015). The combination of practical expertise and one's soft skills, i.e. amalgamation of non-technical and attitudinal competencies, helps professional engineers to become successful (Hissey 2000; Passow and Passow 2017; Azmi et al. 2018; Russo 2016; Leslie 2016). The theoretical knowledge only helps the learners become aware of different functionality of one's ability but the practical execution teaches them how to use these competencies with proficiency in unfavourable situations in real life (Williams et al. 2016).

20.2 Metacognition and Competence in Engineering Education

Metacognition is a higher-order thinking skill or process that makes learners aware of their awareness (Flavell 1979; Wenden 1998). Classified among 21st century skills, metacognition includes self-assessment of the development of critical thinking, communication skills, collaborative practices (Geisinger 2016), indefinite integral problems solving (Sengul and Yasemin 2015), etc. Through metacognition, learners can be more specific about how to monitor and modify the way they think, not only in academics but also in career and life in general (Prytula 2012).

The term competence, first used by Lundberg (1972) in 1970, is a set of demonstrable merits, i.e. the combination of practical and theoretical knowledge, skills, abilities, attitudes and behaviours, awareness and candidness; these competencies enable a person to perform skilfully, make valuable decisions, take effective action and correlate with performance on the job, that can be measured against well accepted standards, and can be improved through training and development (Passow and Passow 2017; Russo 2016; Leslie 2016). Competencies are divided into technical, non-technical and attitudinal/behavioural (Passow and Passow 2017; Russo 2016; Male et al. 2009).

20.3 Literature Review

Metacognition can compensate for the lack of appropriate domain knowledge when metacognitive awareness leads to recognizing the areas of limited understanding, adapting working hypotheses, monitoring thinking, and revisiting early interpretations (Kim and Jeeheon 2013). Even in the new advanced technological era, engineers along with technical skills require employability skills (Robinson 2000), foreign language proficiency and better soft skill proficiency (Gilleard and Gilleard 2002), multilateral and technical and non-technical expertise to deal with professionals in various disciplines (Inman 2006), problem-based learning (Kumar and Natarajan 2007), and competencies expanding beyond countries (Lucena et al. 2008). Engineering educators should focus on developing generic engineering competencies (Male 2010), cognitive competencies (Frank 2012), requirement of the industries (Nair et al. 2009), career aspiration (Itani and Issam 2016; Passow 2013), successful social behaviours (Bakar and Ting 2011), and the current level of skills owned by the staffs (Russo 2016). These competencies should be a part of the curriculum to develop both technical and non-technical skills among engineers (Azmi et al. 2018).

20.4 Objectives

The objectives of the paper are to study how the metacognitive awareness helps Oskar Schindler identify all the limitations and unfavourable conditions. It also focuses on identifying and analysing the competencies used by Oskar Schindler to execute his job and responsibilities.

20.5 Methodology

This qualitative case study is descriptive in nature. The character chosen for the study is Oskar Schindler from the movie Schindler's List. The character of Oskar Schindler is scrutinized for the use of his metacognitive awareness. Use of metacognitive awareness is analysed as per his understanding of the situations, processes and methods. An analysis of his jobs and responsibilities is done to find out the use of generic engineering competencies by him. Generic engineering competencies have been selected from the previous studies and shortlisted according to the need of industries and professional engineers (Russo 2016; Male et al. 2009).

20.6 Results and Analysis

Oskar Schindler used his metacognitive awareness to implement his generic engineering competencies, i.e. the combination of engineering (technical skills) and generic (non-technical/attitudinal skills) competencies to analyse and solve ill structured and ill-defined problems with efficiency (Spielberg et al. 1993). It was his metacognitive awareness that helped him use his competencies. Table 20.1 shows the types of generic engineering competencies, the specific categories of technical, non-technical and attitudinal/behavioural competencies, and the situations where Schindler used his metacognitive awareness to perform his tasks.

Table 20.1 Showing the types of competency, their category, and role of the metacognitive awareness in the use of competencies by Oskar Schindler

Competency	Category	Role of the metacognitive awareness in the use of competencies by Oskar Schindler
Technical	Problem solving	Schindler thought conceptually to define and analyse the problem of money, space, and work force. After evaluating the alternatives, he took money and space from Jews; paid them back in product; hired them as his work force in his factory; and balanced trades-offs (16:37)
	Practical ingenuity	Schindler used his practical knowledge, skills and familiarity with techniques, tools and materials in his factory (1:21:00). As a result, he could get the work done and fulfil his sole customer's needs
	Workplace changes	He recognized the scope for opportunities; made suitable adjustments to open his factory; gathered resources; and utilised them in the best possible way (13:08)
	Manufacturability	He evaluated and improved manufacturability of his product after getting the contracts from the German army (12:23). With his knowledge, awareness, and skills, he got better opportunity to sell products at a reasonably good cost
	Negotiation	Schindler negotiated and convinced his client (the German army) of their needs. He negotiated even with the Jews for money (21:17) and place (20:00). His justification of the need for workforce to make his factory work helped him convince the German army to allot him the required workforce

(continued)

Table 20.1 (continued)

Competency	Category	Role of the metacognitive awareness in the use of competencies by Oskar Schindler
	Design	Schindler used design methodology to gather information plan to set up a factory according to the needs of the German army; generated the idea of starting his factory after checking the feasibility, communicating with Jewish investors and SS officers, and seeking their help and approval
	Liability	After his failure in his previous efforts, he took the risk of starting a new company. Feasibility and execution of this task was maintained by his knowledge of technology, legislation, and standards
	Self-management	Schindler gave priority to his work as well as his personal life. He managed time and punctuality at his workplace. His self-management helped him arrange workers (12:55), provide employment and make profit to earn name for himself. Having established himself as a businessman, ultimately he could save the lives of Jews. Good level of output at his workplace satisfied the need of the German army
	Demeanour	He presented himself as a polite and helpful person, smartly dressed and confident in his apparel and appearance (1:15:20)

(continued)

Table 20.1 (continued)

Competency	Category	Role of the metacognitive awareness in the use of competencies by Oskar Schindler
Non-technical	Communication/Information management	He kept every kind of documents and information with him, whether it be the lists of Jewish workers or any claim funds. Schindler prepared the list of the Jews and maintained the documents to save the lives of the Jews (2:15:54) (2:19:00). He always stayed up to date about the need of the stock, work force, place, and money, indeed. It helped him to understand not only the waste but also the need of it. He clearly presented his agenda before Stern, his accountant and asked him to do accordingly. He could make people listen to him with due respect
	Teamwork	As a good team worker, he trusted, listened and respected other team members irrespective of their identity. He built team cohesion in his factory to manage conflict and work with Jews, SS officers, and even Nazi officers (1:15:54)
	Decision making; analytic skill	Schindler made decisions within the domain of time and knowledge constraints. Using his knowledge and strong analytic power, he successfully planned the commencement of his factory. He arranged for the work force, money, and place. Whatever he knew, he utilized it in the best possible way and worked hard to make things work successfully at any cost (13:30)

(continued)

Table 20.1 (continued)

Competency	Category	Role of the metacognitive awareness in the use of competencies by Oskar Schindler
	Creativity	He thought laterally using his creativity and critical thinking to establish his factory in the most unsuitable conditions and chose the German army as his sole client (12:00)
	Coordinating	Schindler believed in the coordination of the work. He made all discordant elements and situations like limited resources, unfavourable conditions, lack of trained hands (2:15:00), proper permission, resources, etc. fall in place and worked efficiently to make things work in his favour
	Action orientation	Schindler avoided delays and smoothly maintained a sense of urgency, which helped him to save the lives of Jews. He prepared a list of Jews working in his factory, flimsily associated with work force but intentionally associated with human lives (2:15:30)
	Leading	Schindler recruited Jewish team members and managed cooperation to get the job done, motivated and inspired them to work hard, and promised to helped them in return (1:26:47)
	Entrepreneurship	Schindler identified the need of the market, zeroed in on his customer, gathered and utilized all possible resources to focus on the production. Using innovation and commercializing opportunity as an entrepreneur, he successfully switched over from enamel to kitchenware (30:00)

(continued)

Table 20.1 (continued)

Competency	Category	Role of the metacognitive awareness in the use of competencies by Oskar Schindler
	Marketing	His sole customer never seemed to be dissatisfied with the needs and delivery of his products
	Mentoring	Schindler advised and trained his workers and co-workers to continue working despite all limitations and hurdles, which earned them their livelihood as well as lives (23:00)
	Supervising	He supervised the work of his workers by keeping an eye on them from the top of his factory, which was considered to be his favourite place (38:00). He directed the execution of the activities of the workers
	Focus	The centre of Schindler's interests and activity was running the factory, which he maintained by making every possible efforts at multiple levels (39:30)
	Embracing changes	Schindler initiated every possible new approach and technique. It was not that he embraced the choice; in fact, he floated with it as per the technology, capital, initiation, driving force, etc. (1:16:30)
	Life-cycle	Schindler was aware of the life-cycle situation that after the end of the war, his factory would go down, but he still took the risk of starting his project of opening factory and making best products without being worried about the aftermath (12:28)

(continued)

Table 20.1 (continued)

Competency	Category	Role of the metacognitive awareness in the use of competencies by Oskar Schindler
	Workplace politics	He knew how to deal with the workplace politics and social dimensions, both apparent and subtle. He had to be extremely sensitive as he worked in an unusual societal condition, which was quite precarious (11:54)
	Networking	Schindler maintained personal network with workers, co-workers, Stern, organizational and social networks with Nazi and SS officers and with others (31:00)—whosoever could be of any great help to him
	Interdisciplinary	He interacted and traded with people from diverse discipline and professions like Jews for money and space, and Nazi officer and SS officers for infrastructure and work force (2:41:00)
	Honesty	Schindler always demonstrated honesty in his work and made his work related intentions clear to Jews, Nazi officers, SS officers and Stern
	Meeting skills	Officially, Schindler chaired and participated constructively in timely meetings with Nazi officers, contractors and investors (20:00); unofficially with Jewish workers to get help (38:30), maintain work conditions, and manage work force. He involved himself even in community and public debates

(continued)

Table 20.1 (continued)

Competency	Category	Role of the metacognitive awareness in the use of competencies by Oskar Schindler
Attitudinal/behavioural	Concern for others	Schindler thought of the welfare of others in his organization, voluntarily shared information about his factory, ensured fair and liable decisions both for his workers and factory. Facilitating others' contribution, he didn't let anything discard his duties and responsibility (2:45:00)
	Safety	Schindler implemented measures to improve health and safety issues among his workers at all levels, across ages using his best ability not to let his workers go to the camps by justifying their employability at his factory
	Loyalty	Schindler always stayed loyal in representing his company positively to Germans as well as Jews (42:00)
	Ethics	He had the power and justification to kill, cheat or deny the wages or rights but he didn't
	Commitment	Schindler was always committed to doing his best
	Risk taking	Schindler took risks without being afraid of the consequences. He moved beyond, probably all the time. Neither society nor political conditions were favourable for him; still, he took the risk of running the factory
	Up to date	Schindler stayed up to date with current events, contemporary business concepts, and techniques to work efficiently (12:12)

Metacognitive awareness, the knowledge about the person, the task and the strategy, is a learner centred approach. Beyond the curricula, examples of such nature can be referred to, to help the learners think of/visualise situation specific decisions (Conley 2014). Learners can be guided to be more specific in monitoring the task execution and mentoring their own thinking (Williams et al. 2016). Use of such situation specific examples may also allow discussion on the relevance of problem finding before problem solving, seeking close ended as well as open ended solutions, working with the resources available, thinking through the complexities, etc. (Flavell 1979; Prytula 2012; Sengul and Yasemin 2015; Williams et al. 2016). The learners can be made aware that the way they think is liable to be acclimatised when they revisit their earlier interpretations (Kim and Jeeheon 2013).

20.7 Discussion and Implications

Despite any training, technical and domain expertise, Schindler's metacognitive awareness for self-assessment helped him recognize the areas of limited understanding, analyse the adverse situations, solve the complex problems, and be specific about how to monitor and modify the things around him. His metacognitive awareness, known as being aware about awareness, made him use his competencies to solve the problem of money, space, and work force for opening his factory. He believed in 'simple engineering' and used his technical competencies, practical knowledge, skills, ability and familiarity with techniques, tools, materials to analyse and overcome the problems. He used his interpersonal skills, i.e. non-technical competencies to communicate and take situation specific decisions. These competencies compensated for his technical skills in maintaining system safety, building his team cohesively, believing in himself, and making a fortune for himself. His attitudinal/behavioural competencies helped him become aware of the scenario and its demand. He knew how to deal with the needs/demands of German army. His factual and conceptual knowledge formed the foundation of his ability. Application of metacognitive awareness and competencies made him perform his tasks and deal with all the hurdles and **chiselled his behaviour to meet the demands and expectations of his customer**. His positive workplace adaptation led his organization and employees towards improved productivity with higher morale. He remained up to date with current events, contemporary business concepts, and techniques.

20.8 Conclusions

Metacognitive awareness and competencies help in the execution of engineers' aptitudes and abilities. They make them more mindful of what they are doing at their workplace and how these skills and competencies might be used differently in different situations. When it comes to knowledge, there are different kinds of knowledge and different ways of acquiring metacognitive awareness and competencies but it is practical knowledge that often leads to a deeper understanding of a concept through the act of seeing, doing and experiencing.

References

Azmi AN, Kamin Y, Noordin MK (2018) Competencies of engineering graduates: what are the employer's expectations? IJET 7(2.29):519–523. https://doi.org/10.14419/ijet.v7i2.29.13811

Bakar AA, Ting C (2011) Soft skills recommendation systems for IT jobs: a bayesian network approach. In: 2011 3rd conference on data mining and optimization (DMO). IEEE, Selangor Malaysia, pp 82–87. https://doi.org/10.1109/dmo.2011.5976509

Conley DT (2014) Learning strategies as metacognitive factors: a critical review. EPIC, Eugene

Earnest J (2005) ABET engineering technology criteria and competency based engineering education. In: 35th ASEE/IEEE frontiers in education conference. Indianapolis, IN, pp 7–12. https://doi.org/10.1109/fie.2005.1612046

Flavell JA (1979) Meta-cognition and cognitive monitoring: a new area of cognitive development inquiry. Am Psychol 34(10):906–911

Frank M (2012) Engineering systems thinking: cognitive competencies of successful systems engineers. In: Dagli CH (ed) New challenges in systems engineering and architecting conference on systems engineering research (CSER) 2012, Procedia computer science, vol 8. Elsevier, Missouri, pp 273–278. https://doi.org/10.1016/j.procs.2012.01.057

Geisinger KF (2016) 21st Century skills: what are they and how do we assess them? Appl Measur Educ 29(4):245–249. https://doi.org/10.1080/08957347.2016.1209207

Gilleard J, Gilleard JD (2002) Developing cross-cultural communication skills. JPI 128(4):187–200. https://doi.org/10.1061/(ASCE)1052-3928(2002)128

Hissey TW (2000) Education and careers 2000: enhanced skills for engineers. Proc IEEE 88(8):1367–1370. https://doi.org/10.1109/5.880089

Inman W (2006) What are you talking about? Want to see your proposals sail past management? Stop sounding like an engineer. Ind Eng ABI/INFORM Glo 36(1)

Itani M, Issam S (2016) Engineering students' perception of soft skills, industry expectations, and career aspirations. JPI 142(1):04015005. https://doi.org/10.1061/(asce)ei.1943-5541.0000247

Kim M, Jeeheon R (2013) The development and implementation of a web-based formative peer assessment system for enhancing students' metacognitive awareness and performance in ill-structured tasks. Education Tech Research Dev 61(4):549–561

Kumar M, Natarajan U (2007) A problem-based learning model: showcasing an educational paradigm shift. Curr J 18(1):89–102. https://doi.org/10.1080/09585170701292216

Leslie C (2016) Engineering competency model. ASEE Ann Conf Exposi Proc. https://doi.org/10.18260/p.26627

Lucena J, Downey G, Jesiek B, Elber S (2008) Competencies beyond countries: the re-organization of engineering education in the United States, Europe, and Latin America. JEE 97(4):433–447

Lundberg CC (1972) Planning the executive development program. Calif Manag Rev 15(1):10–15. https://doi.org/10.2307/41164394

Male SA (2010) Generic engineering competencies: a review and modelling approach. ERP 37(1):25–51

Male SA, Bush MB, Chapman ES (2009) Identification of competencies required by engineers graduating in Australia. In: Kestell C, Grainger S, Cheung J (eds) 20th conference of the AAEE: engineering the curriculum. School of Mechanical Engineering, the University of Adelaide, The University of Adelaide, pp 882–887

Nair CS, Patil A, Mertova P (2009) Re-engineering graduate skills: a case study. EJEE 34(2):131–139

Passow HJ (2013) Which ABET competencies do engineering graduates find most important in their work? JEE 101(1):95–118

Passow HJ, Passow CH (2017) What competencies should undergraduate engineering programs emphasize? A systematic review. JEE 106(3):475–526

Prytula MP (2012) Teacher metacognition within the professional learning community. IES 5(4):112–121. https://doi.org/10.5539/ies.v5n4p112

Robinson JP (2000) What are employability skills? Alabama Cooper Extension Sys 5(3):1–3

Russo D (2016) Competency measurement model. In: European conference on quality in official statistics (Q2016). Madrid, 31 May–3 June 2016

Sengul S, Yasemin K (2015) Meta-cognitive aspects of solving indefinite integral problems. In: 7th World conference on educational sciences, (WCES-2015) Procedia-Social and Behavioral Sciences. Elsevier, Athens-Greece, pp 622–629. https://doi.org/10.1016/j.sbspro.2015.07.205

Spielberg S, Molen GR, Lustig B, Spielberg S (1993) Schindler's list. Universal Pictures, United States

Wenden AL (1998) Metacognitive knowledge and language learning. Appl Linguist 19(4):515–537

Williams SA, Morelock J, Matusovich H, Cunningham P (2016) Lessons in transfer: better understanding of engineering students' metacognitive development. Paper presented at the frontiers in education conference, Erie, PA

21

Indian Engineering Graduated: Development of Cross-Cultural Competence

Pushp Lata

Abstract Multiculturalism being a global reality of the present world, it has become imperative to develop Cross-Cultural Competence (CCC) to become successful as a professional and function effectively as an individual in personal and social interactions. Considering the fact that India with 60 percent of its population being under 25 years of age, it must educate its young generation to succeed in a globalized world. Another vital fact that endorses the need for change is revealed in a study, which states that about 20–33% of the 1.5 million engineers who pass out every year in India, run the risk of not getting a job at all. Among the multiple reasons for their being unemployable, one of the strongest is their lack of exposure to "other" cultures and their nuances that makes it hard for the leading MNCs to hire them. Keeping these facts in mind the present study argues the need for capacity building of engineering graduates through cross cultural skills. This paper recommends a course on Cross Cultural Skills to be offered to engineering students in order to inculcate in them the desired global competency among them. Besides this, the paper also discusses the different types of activities and tasks that are done in Cross Cultural Skills (CCS) classroom at BITS Pilani with a view to grooming students as effective University ambassadors. It also shares the feedback given by students to highlight the need for developing cross cultural skills and the success of the training during the course.

Keywords Engineering education · Global competence · Cross cultural skills

P. Lata (✉)
Department of Humanities and Social Sciences, Birla Institute of Technology and Science, Pilani-Campus, Pilani 333031, RJ, India
e-mail: plat@pilani.bits-pilani.ac.in

21.1 Introduction

In the changing times of digital era, engineers and technocrats are expected to be globally competent, which means the ability to work knowledgeably and live comfortably in a transnational engineering environment and global society (Lohmann et al. 2006). The most significant underpinning for the need of developing intercultural competence (IC) is the multicultural reality that we live in today. Our society has become "increasingly international, with abundant intercultural encounters through tourism, trade, diplomacy, exchange and the mobility of people" (Larzen 2005; Weaver 1993; Woodin 2001; Lata 2017). Consequently, cultural diversity puts a demand on our contemporary society for timely education-based responses for a change in perspective concerning 'the others' and 'ourselves' (Cheh 2015). Daft and Dorothy too seem to endorse similar idea suggesting that if we look at the business world around where the students of today might operate in future, we find that there is a drastic change in the desirable traits of professionals and managers. In the post globalization era the market has now crossed trade borders of the nations bringing a shift towards a more integrated and interdependent world economy. As part of a global workforce, the need for personal learning and growth are crucial while managing business affairs with a hybrid team (Daft and Marcic 2006). The recent innovations and development in information and communication technology have changed the way we communicate and do businesses. This change has further prompted a transformation in the role of engineers in the society and, consequently, the nature of engineering practice (Loftus 2003). Further, the Hart Research Associates study too seems to augment the above stated fact when their report reveals that 78% of the surveyed employers unambiguously emphasized on the need for all the students gaining intercultural skills (Hart Research Associates 2015). Therefore, in order to stay relevant, higher education institutions must take charge of their internationalization and produce graduates who will excel in the global work (Fellows 2014). So, for making their graduates culturally competent, it becomes imperative for universities across the globe to explore effective and suitable teaching and learning approaches in a cross-cultural education environment.

Similarly, the universities in India too need to incorporate this international preparation into engineering curricula by introducing courses related to Intercultural Communication or Cross Cultural Skills and prepare their engineers for transnational practice and help them become part of global society. As of February 2017, there are 789 universities, 37,204 colleges and 11,443 stand-alone institutions in India, as per the UGC website (https://www.ugc.ac.in/oldpdf/Consolidated%20list%20of%20All%20Universities) and there are 4,282 engineering institutes in India approved by AICTE (AICTE approved Institutes for the academic year 2017) and a preliminary study suggests that except for a couple of IITs (Indian Institutes of Technology) and a few management schools, no course or training is being provided in these academic institutes to make their graduates aware of intercultural skills that will enable them to adapt and work smoothly in a multicultural setting. The author also conducted a study on 226 university students and the results of the survey reflected that 64.1% of the respondents strongly felt the need for such kind of training in cross cultural skills, which has spurted due to the increased heterogeneity both in educational setting and the professional work environment (Lata et al. 2018).

The ensuing section entails briefly about how the course on Cross Cultural Skills develops intercultural understanding among students.

21.2 Methodology

For this study the course plan of CCS Course offered at BITS Pilani and the assignments and activities that are done during the semester, have been explored and discussed to show how a structured course at the university level can help in training the students for developing global competence. Besides, the data base of the course and the feedback given by the students have also been referred to for supporting the arguments positioned in the paper.

21.3 Cross Cultural Skills Course at BITS, Pilani

Cross Cultural Skills is being offered as an elective course at BITS Pilani to its engineering students since 2013. The objective of the course is to help students understand the need of cross-cultural skills and provide insights into intercultural communication by familiarizing them as to how to deal with ethnocentrism, personal biases and self-reference, which may hinder their appreciation of the cultural diversity prevalent in communities worldwide and operate successfully. It aims at helping the students discover the kinds of barriers and challenges individuals confront while working beyond cultural boundaries. It also aims at training and imparting those necessary skills so that they know how to overcome culture shock and successfully negotiate the cultural alienness, adapt its nuances and integrate cultural heterogeneities while working in a multicultural team. Besides, it provides a comprehensive understanding of how and why cultural groups differ and how their impact determines success or failure of international businesses.

The following table describes the course content, the teaching methodology adopted and the kind of assignments are conducted in the course (Table 21.1).

Students are also taken to the nearby villages for helping them develop a practical understanding into the similarities and dissimilarities in people, places and their activities when they belong to different caste, class, religion and culture. They are asked to create a fact sheet about their field visit in which they are also asked to share at least two interesting incidents/anecdotes so as to share with others their minute findings and observations during survey along with the shaping factors they could comprehend during their visit.

Besides these hands on experiences, discussions are held on the "Global Leadership and Organizational Behavior Effectiveness" (GLOBE) by House (House et al.

Table 21.1 Details of cross cultural skills course at BITS Pilani

Course contents	Teaching methods	Assignments
• Edward Hall's model • Geert Hofstede's cultural dimensions • Richard Lewis' cultural dimensions • Milton Bennett's developmental model of intercultural sensitivity (DMIS), (Bennett 2002) • Fons Trompenaars' seven cultural dimensions(Fons and Turner 1997) • Kluckhohn and Strodtbeck's dimensions • Robert J. House's global leadership and organizational behavior effectiveness (GLOBE) • Critical issue in intercultural communication: practice challenges and barriers • Managing global teams	• Lectures • Pair work • Group work • Field work • Videos • Discussion over films and literary works	• Creating cultural bag—knowing thyself and thy culture • Fact file—knowing the other based on field work • Role plays for cross culture communication barriers • Team presentations • Case study discussions • Concept JAM • Creative project/special project

1991) and Kluckhohn and Strodtbeck's dimensions (Hills 2002) and their influence on workplace policies. Several case studies are taken up and discussed at length. Business case studies such as Walt Disney in Paris, Wal-Mart in Germany, Daimler-Benz Chrysler, Coca Cola in India, McDonald in India, Snapdeal and Flipkart merger, Shree Renuka Sugars buyout of Brazilian Companies, Bharti Airtel–Zain and so on and so forth, are also taken up for analysis and discussion (Madhavan 2011).

For group assignment students are divided into different groups consisting of five students each and they are given different situations for performing the roles of different stakeholders in a business situation like some human resource issue, promotion decision, negotiation, marketing, and they are asked to discuss and create the whole dialogue and perform for about 10 min. After two- week time of preparation, the groups are asked to perform. This assignment helps students conduct in-depth research, decide on the situations, characters and create dialogues and finally perform as a team. This involves students in using their critical thinking and higher order thinking skills to analyze, evaluate and create as suggested by Revised Blooms taxonomy grid (Airasian et al. 2001).

The course also includes a choice of doing either Special Project Assignment or a Creative Project. Special Project Assignment allows the students to choose their own question in relation to the cultures and then they are supposed to write a short, analytical paper in 3,000 words. They are asked not to make that their question, issue or problem too general. They are supposed to come up with a thesis statement which summarizes their argument about the question, theme or issue they grapple with in their paper. All their claims about the cultures are to be supported by evidence or case studies (using proper citation methods).

Alternatively, students may opt for doing creative project. In this project a lot of flexibility is given. They can shoot a film or make a video or create a photo collage or write a poem or prepare a short skit or take up a creative writing project. The fundamental requirement is that the creative project should relate to the themes of the course. They are to pick up at least two cultures as subject of study here. Again, it cannot be too general. For instance, they cannot just string together a few photographs titled Alienation. The creative project should reflect a thoughtful engagement with at least two cultures. In fact, they are also required to pose a question in relation to any culture (at least two) for their creative project, and attempt to explore it through the form they have chosen. They are required to follow the guidelines for the academic paper to ascertain a sense of rigorous engagement with the project. They are asked to submit a 500 word document, which describes their project, the questions raised from the cultures they choose as well as give them room for their creative expression, which culminates in their final submission.

In the last semester for the first time this assignment was introduced; 33 out of 90 students opted for creative project under which 13 students composed poems, 7 students shot videos, 6 students drew paintings, 3 tried writing skit and 4 students used photography to present their understanding into two different cultures. Based on the data collected through google form 86.4% appreciated the flexibility and the space, which enabled them to use the mode through which they could apply their knowledge of cultures.

Hence, the CCS course provides multifarious practical assignments to develop cultural intelligence to students and with the enhanced exposure into cultural revelations they can systematically deal with stereotypical and ethnocentric behaviours, which usually become major hurdles in intercultural communication environment.

21.4 Results

The database created for the course reflects that so far 448 students have done this course since 2013; only 11.9% had some kind of international exposure but rest 88.1% did not have any exposure at the outset of the course. During the first class of this course every semester, the students are asked about the problems they face while interacting with people from other cultures. Both the sets of students with or without international exposure have consensus regarding the fact that they encounter unnecessary conflicts and a feeling of otherness during interactions. The students with international exposure hence, show better conviction and clarity about the significance of doing a course like CCS. Students have started realizing that with Global Businesses and international work assignments are increasing at a breakneck pace in multinational corporations and as future professionals they need to develop global skill sets. The feedback collected at the end of the course reflects that 85% of the students feel that after doing the course they have become more comfortable in interacting with people from other cultures. This reflects how the training helped the students in developing their interpersonal skills in multicultural environment. 76.7% of the students felt that course has developed better skills for working in multicultural teams. Since they were working in hybrid teams and collaborating on their assignments throughout the course, they could align both their communication and working style with that of their team members. Students after completing the course when they go for internships in companies they come and share how the training helped them.

21.5 Conclusions

In brief, it is time that Indian engineering colleges, institutes and universities recognized the need for providing enhanced exposure to their students into the cultural complexities of the professional world so as to make their students globally competent. Considering the example of the Cross Cultural Skills course at BITS Pilani, it can be suggested that a well-defined course on Intercultural communication or cross cultural skills can enhance their intercultural competence and help the nation reap the benefit of its demographic dividend. Further quantitative and qualitative studies can be done the effectiveness of the Cross Cultural Skills in the development of intercultural sensitivity in students.

References

AICTE approved Institutes for the academic year: 2017–2018: AICTE. (2017) https://www.aicte-india.org/downloads/Get_Institute_Id.pdf. Last accessed 11 May 2019

Airasian PW et al (2001) A taxonomy for learning, teaching, and assessing. In: Anderson LW, Krathwohl DR (eds) A revision of bloom's taxonomy of educational objectives. Longman, New York

Bennett MJ (2002) Becoming interculturally competent. In: Wurzel JS (ed) Toward multiculturalism: a reader in multicultural education. Intercultural Resource Corporation, Newton, MA

Cheh O (2015) Intercultural competence in foreign language teaching and learning in Indian institutes. Unpublished thesis, JNU. http://hdl.handle.net/10603/117895. Last accessed 16 July 2018

Daft RL, Marcic D (2006) Understanding management. Cengage Learning, Delhi

Fellows KL et al (2014) What's your CQ? A thought leadership exploration of cultural intelligence in contemporary institutions of higher learning

Fons T, Turner CH (1997) Riding the waves of culture: understanding cultural diversity in business. Nicholas Brealey, London

Hart Research Associates (2015) Falling short? College learning and career success. A Report on Online Survey conducted by Association of American Colleges and Universities

Hills MD (2002) Kluckhohn and Strodtbeck's values orientation theory. Online Readings in Psychol Culture 4(4). Last accessed 21 Aug 2019. https://doi.org/10.9707/2307-0919.1040

House et al (1991) Cultural influences on leadership and organizations. J Adv Global Leader 1:171–233

Larzen E (2005) In pursuit of an intercultural dimension in EFL-teaching: exploring cognitions among Finland-Swedish Comprehensive School Teachers. Åbo Akademi University Press, Faculty of Education

Lata P (2017) Culturally responsive pedagogy for effective teaching of communication skills. In: Govil JN et al (eds) Innovation and education. Studium Press India, New Delhi, pp 71–85

Lata P, Iyenger RK, Yadav S (2018) Intercultural competence of Indian engineering students: an assessment and need for development training. In: Proceedings of ISERD international conference. Oslo, Norway, pp 6–10

Loftus M (2003) A new era. Prism 12(8):26–29

Lohmann JR, Howard AR, Hoey JJ (2006) Defining, developing and assessing global competence in engineers. EJEE 31(1):119–131. https://doi.org/10.1080/03043790500429906

Madhavan S (2011) Cross-cultural management: concepts and cases, 2nd edn. Oxford University Press, New Delhi

UGC Report on Education in India. https://www.ugc.ac.in/oldpdf/Consolidated%20list%20of%20All%20Universities. Last accessed 10 Aug 2019

Weaver G (1993) Understanding and coping with cross-cultural adjustment stress. In: Paige RM (ed) Education for the intercultural experience. Intercultural Press, Yarmouth, Maine, pp 137–168

Woodin J (2001) Tandem learning as an intercultural activity. In: Byram M, Nichols A, Stevens D (eds) Developing intercultural competence in practice. Multilingual Matters, Clevedon, England, pp 189–202

22

ITLC in India and Germany

Luise Mayer, Elena Stasewitsch, and Simone Kauffeld

Abstract An innovative teaching and learning climate (ITLC) describes the positive attitudes of teachers and students towards developing and applying innovation, i.e., new ways of thinking and practices at their university. ITLC is assumed to positively affect many aspects of teaching and learning outcomes, e.g., sustainable thinking. Contrary to the hypothesis, the German university TU Braunschweig ($N = 394$) did not score higher in ITLC than BITS Pilani ($N = 484$). Sustainable thinking correlated with ITLC. This is the first study to provide insights into how ITLC relates to sustainable thinking. Thus, the results supplement existing research on sustainable thinking.

Keywords Innovative climate · Higher education · Sustainable thinking

22.1 Introduction

Across the world, innovating and enabling sustainable development are constantly gaining importance. As "centers of knowledge generation and sharing" (Nejati et al. 2011) universities bear the responsibility for these societal tasks (Nejati et al. 2011). Their challenge is to prepare future specialists and executives all over the world. However, even universities themselves lack innovations (Blume et al. 2015; Wals and Jickling 2002) and the ability to enable sustainable development on campus (Velazquez et al. 2005).

One speaks of *innovation* when a group beneficially (Nonaka 1994) applies (O'Sullivan and Dooley 2008) an idea that is new to them (Rogers, 2010). In this research, innovation includes innovative teaching and learning, and sustainable development. A high *innovative climate* describes an environment that is particularly conducive to innovations: Individuals share (Van der Vegt et al. 2005) positive attitudes (Alisch et al. 2013) towards new ideas and developments within their group (Van

L. Mayer (✉) · E. Stasewitsch · S. Kauffeld
Division of Industrial/Organizational and Social Psychology, Institute of Psychology, Technische Universität Braunschweig, Spielmannstr. 19, Braunschweig 38106, Germany
e-mail: luise.mayer@magenta.de

der Vegt et al. 2005). Mutual appreciation and intensive communication are characteristic (Stasewitsch and Kauffeld (in press)). A high innovative climate leads to innovative behavior (Yu et al. 2013) and therefore, to innovations (Scott and Bruce 1994). This connection was found for educational institutions as well (Sagnak 2012). We use the more university-specific term *innovative teaching and learning climate* (ITLC) in this research.

India has the third largest higher education system in the world (Gupta and Gupta 2012). However, its reputation for education and research is relatively poor across the globe (Sharma and Sharma 2015). Probably related to this is the low employability (Sharma and Sharma 2015). Practical methods (Aymans et al. 2017; Büth et al. 2017) like internships, career counselling, and international projects (Sharma and Sharma 2015) are rarely applied in India. Whereas in Germany, innovative teaching is widespread (Kauffeld and Othmer 2019). Therefore, we propose:

- **Hypothesis (H) 1.1** Didactic and technical innovations (mobile, individualized, research-orientated learning) are applied less frequently in Indian than in German universities.
- **H 1.2** Indian universities have a lower innovative climate than German universities.

The comparison of an Indian university, Birla Institute of Technology and Science, Pilani, India (BITS Pilani) with an German university, Technische Universität Braunschweig (TU Braunschweig) should identify factors inhibiting innovations on BITS Pilani campus. Building on this, we aim to provide recommendations to BITS Pilani for actions to improve the ITLC. This can contribute to replacing outdated teaching and learning methods with innovative ones.

Sustainable development defines the "development that meets the needs of the present without compromising the ability of future generations to meet their own needs" (Brundtland et al. 1987, p. 37). It can be regarded as innovation even at German universities (Richter and Hunecke 2017) and as a fairly new concept in developed countries in general (Meadowcroft 2000). Consequently, sustainable thinking and acting can be understood as innovative behavior. That is why we assume that innovative climate can promote sustainable thinking. Our assumption is also supported by the fact that sustainable development at universities fails, among other things, due to a reluctance to change and poor communication (Velazquez et al. 2005). In a high ITLC, the opposite is the case; the members of the organization welcome new ideas (Scott and Bruce 1994) and have intensive relations in general (Moolenaar et al. 2011). That means ITLC tackles exactly the obstacles of sustainable development at universities.

- **H 2.1** ITLC correlates positively with sustainable thinking.

At the moment, knowledge about sustainable development (Zsóka et al. 2013) and the status, i.e., the reputation of acting sustainably (Griskevicius et al. 2010) are considered to be the most important influencing factors on sustainable thinking and acting.

- **H 2.2** Knowledge about sustainable development correlates positively with sustainable thinking.
- **H 2.3** The status of acting sustainably correlates positively with sustainable thinking.

This research aims to determine factors inhibiting and encouraging innovations at BITS Pilani. When purposefully increasing the ITLC, university members are more likely to generate innovations. This is shown by the innovation sustainable thinking. To classify the strength of the relation between ITLC and sustainable thinking, it is compared with that of known factors influencing sustainable thinking.

22.2 Method

BITS Pilani members, students and teachers took part in this research. For comparison, previously collected data (Stasewitsch and Kauffeld (in press)) from Germany (TU Braunschweig) were used. Table 22.1 shows the sample composition in terms of occupation, age, and gender.

For measuring the ITLC *Inno-Teaching* (or respectively, its original German version *Inno-Lehre* (Stasewitsch and Kauffeld (in press)) was applied. It contains twelve scales including three about didactic and technical innovations, namely *Mobile, Individualized* and *Research-orientated Learning*. Additionally, sustainable thinking (Biasutti and Frate 2017), knowledge about sustainable development (Azapagic et al. 2005), and the status of acting sustainably (Richins 1994) were measured (see Table 22.2).

Table 22.1 Sample composition

Sample 1 (BITS Pilani)	Sample 2 (TU Braunschweig)	Sample 3 (BITS Pilani)
$n^t = 67$ Ø = 36.2 (*SD* = 10.5) 88 ♂ & 12% ♀	$n^t = 205$ Ø = 39.4 (*SD* = 11.2) 62 ♂ & 32% ♀	
$n^s = 417$ Ø = 20.2 (*SD* = 1.9) 86 ♂ & 13% ♀	$n^s = 189$ Ø = 24.8 (*SD* = 3.2) 56 ♂ & 41% ♀	$n^s = 221$ Ø = 20.0 (*SD* = 1.8) 87 ♂ & 13% ♀

Notes n = number of participants, t = teachers, s = students, Ø = average age, SD = standard deviation

Table 22.2 Applied scales

Measured construct	Applied scales	Subscales (number of items)
Innovative teaching and learning climate	Inno-Lehre and Inno-Teaching (Stasewitsch and Kauffeld (in press))	• Appreciation (3) • Relationship quality (3) • Networking in teaching (3) • Activating leadership (3) • Activating support (3) • Dealing with mistakes (2) • Reflection: Teachers (3) • Reflection: Students (2) • Aims in teaching (2) • Qualification in teaching (3) • D. & t. i.: Mobile learning (3) • D. & t. i.: Individualized learning (2) • D. & t. i.: Research orientated learning (3)
Sustainable thinking	Attitudes towards sustainable development (Biasutti and Frate 2017)	• Environment (5) • Economy (5) • Society (5) • Education (5)
Knowledge about sustainable development	Knowledge about sustainable development (Azapagic et al. 2005)	• Environmental issues (14) • Environmental legislation, policy, and standards (7) • Environmental tools, technologies and approaches (12) • Sustainable development (12)
Status of acting sustainably	Possession rating scale (Richins 1994), adapted to acting sustainably instead of owning possessions	(3)

Notes D. & t. i. = Didactic and technical innovations

22.3 Results

22.3.1 Innovative Teaching and Learning Climate at BITS Pilani

We rejected Hypotheses 1.1 and 1.2 as BITS Pilani (Sample 1) had higher values than TU Braunschweig (Sample 2) in the analyzed ITLC scales (see Table 22.3).

Table 22.3 Comparison of ITLC between at BITS Pilani and TU Braunschweig

Scale	*BITS Pilani*		*TU Braunschweig*	
	M	SD	M	SD
Mobile learning	2.71	0.75	1.74	0.92
Individualized learning	2.18	0.95	1.45	0.97
Research orientated learning	2.63	0.84	2.14	0.99
ITLC total	75.30	16.96	60.99	19.54

Notes M = Mean score, *SD* = standard deviation. *ITCL* total is the sum of all 32 ITCL items (see Table 22.2)

22.3.2 Sustainable Thinking and Innovative Teaching and Learning Climate

The strength of an association can be estimated by Pearson's r (Cohen and Manion 1980). According to Hypothesis 2.1 and 2.3, sustainable thinking correlated significantly to a small extent with ITLC ($r = .22, p = .00$) and status of acting sustainably ($r = .17, p = .01$). However, sustainable thinking did not correlate with knowledge about sustainable development ($r = -.02, p = .41$).

22.4 Discussion and Outlook

22.4.1 Innovative Teaching and Learning Climate at BITS Pilani

Discussion. Contrary to expectations, BITS Pilani scored higher than TU Braunschweig, not only in the selected scales but also in *Inno-Teaching* as a whole. First of all, a problem that has occurred in the course of this research should be noted. The Hypotheses had been derived from differences between India and Germany, e.g. differences in the reputation in research and education (see 1). These were then only operationalized as BITS Pilani and TU Braunschweig, i.e., only tested at one university per country. This limits the generalizability of the results.

The high score of BITS Pilani may have two causes which are both considered in this section. Either the ITLC at the TU Braunschweig is not more pronounced compared to BITS Pilani, or the ITLC at BITS Pilani is lower than the one TU Braunschweig, but our questionnaire study was unable to capture this.

On the one hand, BITS Pilani possibly achieved a higher ITLC score in our survey despite a indeed lower ITLC than TU Braunschweig because culture influences the response behavior. In collectivist cultures (e.g., India) the group has priority instead of the individual as opposed to individualistic cultures (e.g., Germany) where the individual is in the focus (Triandis 1995). In collectivist cultures, it is crucial to make a good impression when in contact with strangers (Triandis 1995). Thus, Indian participants in a questionnaire study tend to respond in a socially desirable way, i.e., as they believe that they correspond to the social or cultural norm to present themselves or their group positively (Bernardi 2006). Not only culture but also different experiences influence the response behavior of individuals. The anchoring effect says individuals tend to compare things with something else that is present to them at that moment (Tversky and Kahneman 1974). Consequently, BITS Pilani members compare their experience at a private educational institution to financially worse off public educational institutions in their country. Both, social desirability and a downward comparison of the participants, might have lead to higher values regarding the ITLC of BITS Pilani.

On the other hand, the high ITLC score of BITS Pilani could result from the actual high ITLC. Perhaps the Indian education system, or at least its ITLC, is better than its reputation, or BITS Pilani sets an excellent example for Indian universities.

Similar to TU Braunschweig (www.tu-braunschweig.de/teach4tu), at BITS Pilani, a central institute provides further training for teachers and supports innovative teaching (www.bits-pilani.ac.in/tlc/Pilani/index.html). There is broad support for entrepreneurship, including financial support (www.bits-pilani.ac.in/pilani/centerforentrepreneurial/Home), events and courses (Kumar and Jain 2015). Consequently, many alumni found start-ups (Kumar and Jain 2015). An advantage of India compared to Germany is the flexibility (Banerjee 2008). This flexibility allows the implementation of innovations and new ideas in the teaching.

Our study shows a high ITLC at an Indian university. From this result one can conclude the recommendation for German institutes to take a closer look at the "Indian way of teaching" with the aim of finding ideas, such as a greater support for entrepreneurship and enhancing flexibility.

Outlook. To what extent social desirability has indeed played a role in the comparison between BITS Pilani and TU Braunschweig can only be analyzed with further questioning by a scale interrogating social desirability. In that way, the cross-cultural comparison would more likely find out what makes one university better than another and could enable universities from different cultures to learn from each other. For the enabling of a more generalizable cross-cultural comparison in future research, it would also be helpful to record and statistically monitor the characteristics of students and universities as well as to include a private university in Germany and a public university in India. The questioning of students and teachers with experience abroad who know both universities would limit the anchoring effect and therefore increase the comparability of the results.

22.4.2 Sustainable Thinking and Innovative Teaching and Learning Climate

Discussion. The association between sustainable thinking and ITLC has been found for the first time. Moreover, it was the strongest correlation found compared with known influencing factors like knowledge about sustainable development and status of acting sustainably. Our research provides the first indications that besides teaching pure knowledge about sustainable development and presenting sustainability as *cool*, sustainable thinking can also be enhanced by ITLC. Thus, so far, an important factor in education for sustainable development would have been neglected.

However, as in all correlative studies, there is one significant limitation that the causal relations remain unclear. An ITLC might indeed enhance sustainable thinking. Alternatively, the two constructs could influence each other. Or, a third variable such as high social status might be associated with both the factors.

Outlook. This link should be pursued further. Longitudinal studies would be valuable to examine the cross-lagged associations between sustainable thinking and an ITLC.

Acknowledgements The research for this paper is part of the project "JInGEL—Joint Indo German Experience Lab", a joint project between Technische Universität Braunschweig and Birla Institute of Technology and Science Pilani, and is funded by the German Academic Exchange Service (DAAD) under grant number 57219215. The authors are thankful for the funding and the support. Elena Stasewitsch and Simone Kauffeld gratefully acknowledge support from the German Ministry of Education and Research under grant number 01PL17043.

References

Alisch K, Winter E, Arentzen U (2013) Gabler Wirtschafts Lexikon. Springer

Aymans S, Friese H, Kauffeld S (2017) Bildung ohne Grenzen. Intercul J: Online-Zeitschrift für interkulturelle Studien 17:61–81

Azapagic A, Perdan S, Shallcross D (2005) How much do engineering students know about sustainable development? J Eng Educ 30:1–19. https://doi.org/10.1080/03043790512331313804

Banerjee S (2008) Dimensions of Indian culture, core cultural values and marketing implications. Int J Cross Cult Manag 15(4):367–378. https://doi.org/10.1108/13527600810914157

Bernardi RA (2006) Associations between Hofstede's cultural constructs and social desirability response bias. J Bus Ethics 65:43–53. https://doi.org/10.1007/s10551-005-5353-0

Biasutti M, Frate S (2017) A validity and reliability study of the attitudes toward sustainable development scale. Environ Educ Res 23:214–230. https://doi.org/10.1080/13504622.2016.1146660

BITS Pilani: Centre for Entrepreneurial Leadership. www.bits-pilani.ac.in/pilani/centerforentrepreneurial/Home. Last accessed 6 June 2019

Blume S, Madanchi N, Böhme S, Posselt G, Thiede S, Herrmann C (2015) Die Lernfabrik—Research-based learning for sustainable production engineering. Procedia CIRP. https://doi.org/10.1016/j.procir.2015.02.113

Brundtland GH, Khalid M, Agnelli S, Al-Athel S, Chidzero B (1987) Our common future. New York

BITS Pilani: Teaching and Learning Centre. www.bits-pilani.ac.in/tlc/Pilani/index.html. Last accessed 6 June 2019

Büth L, Bhakar V, Sihag N, Posselt G, Böhme S, Sangwan KS, Herrmann C (2017) Bridging the qualification gap between academia and industry in India. Procedia Manufact 9:275–282. https://doi.org/10.1016/j.promfg.2017.04.009

Cohen L, Manion L (1980) Research methods in education, 6th edn. Croom Helm, London (1980). https://doi.org/10.4324/9780203720967

Griskevicius V, Tybur JM, van den Bergh B (2010) Going green to be seen: status, reputation, and conspicuous conservation. J Pers Soc Psychol 98:392. https://doi.org/10.1037/a0017346
Gupta D, Gupta N (2012) Higher education in India: structure, statistics and challenges. J Edu Pract 3:17–24
Kauffeld S, Othmer J (2019) 12.1 Zielsetzung: Forschung für die Praxis–176 12.2 Forschendes Lernen in der Hochschulpraxis–176 12.3 Fazit: forschendes Lernen als studiengangprägendes Konzept–186. Handbuch Innovative Lehre 175. https://doi.org/10.1007/978-3-658-22797-5
Kumar A, Jain A (2015) Factors affecting entrepreneurial intent among students: a case study of BITS Pilani. In: 11th biennial conference on entrepreneurship. EDII, Gandhinagar (2015)
Meadowcroft J (2000) Sustainable development: a new (ish) idea for a new century? Polit Stud 48:370–387. https://doi.org/10.1111/1467-9248.00265
Moolenaar NM, Daly AJ, Sleegers PJC (2011) Ties with potential. Teach Coll Rcco 113:1983–2017
Nejati M, Shafaei A, Salamzadeh Y, Daraei M (2011) Corporate social responsibility and universities. Afr J Bus Manage. https://doi.org/10.5897/AJBM2015.7993
Nonaka I (1994) A dynamic theory of organizational knowledge creation. Organ Sci 5:14–37
O'Sullivan D, Dooley L (2008) Applying innovation. Sage Publications
Richins ML (1994) Valuing things. J Consum Res. https://doi.org/10.1086/209414
Richter N, Hunecke M (2017) Die Verbreitung von Nachhaltigkeit als soziale Innovation an Hochschulen aus Sicht studentischer Nachhaltigkeitsinitiativen. In: CSR und Wirtschaftspsychologie. Springer, pp 297–308. https://doi.org/10.1007/978-3-662-52746-7_17
Sagnak M (2012) The empowering leadership and teachers innovative behavior: the mediating role of innovation climate. Afr J Bus Manage 6:1635–1641. https://doi.org/10.5897/AJBM11.2162
Scott SG, Bruce RA (1994) Determinants of innovative behavior. Acad Manag J 37:580–607. https://doi.org/10.5465/256701
Sharma S, Sharma P (2015) Indian higher education system. Electro J Inclus Edu 3:1–4
Stasewitsch E, Kauffeld S (in press) Der Inno-Lehre: Ein Inventar zur Messung von innovativem Lehr-Lernklima an Hochschulen. Diagnostica
Stasewitsch E, Kauffeld S (in prep.) The Inno-Teaching: a cross-cultural validation
Triandis HC (1995) Individualism and collectivism. Westview Press, Bouder, CO
TU Braunschweig: teach4TU. https://www.tu-braunschweig.de/teach4tu. Last accessed 6 June 2019
Tversky A, Kahneman D (1974) Judgment under uncertainty: heuristics and biases. Science 185:1124–1131
Van der Vegt, Gerben S, van de Vliert E (2005) Huang X location-level links between diversity and innovative climate depend on national power distance. Acad Manage J 48:1171–1182. https://doi.org/10.5465/amj.2005.19573116
Velazquez L, Munguia N, Sanchez M (2005) Deterring sustainability in higher education institutions. Int J Sus in Higher Ed 6:383–391. https://doi.org/10.1108/14676370510623865
Wals AEJ, Jickling B (2002) "Sustainability" in higher education. Int J Sus in Higher Ed. https://doi.org/10.1108/14676370210434688
Yu C, Yu T-F, Yu C-C (2013) Knowledge sharing, organizational climate, and innovative behavior: a cross-level analysis of effects. Soc Behav Personal Int J 41:143–156. https://doi.org/10.2224/sbp.2013.41.1.143
Zsóka Á, Szerényi ZM, Széchy A, Kocsis T (2013) Greening due to environmental education? J Clean Prod 48:126–138. https://doi.org/10.1016/j.jclepro.2012.11.030

23

Engineering Education: Impact of Self-Efficacy

Abhijith Venugopal, Rajni Singh, and Devika

Abstract Self-efficacy, one's perception and self-belief in their abilities to do a particular task successfully, plays a crucial part in engineering education. It helps students enhance and strengthen their learning skills. The recent research has added multifaceted and integrated self-efficacy as the main factor of learning, both theoretical and practical. When placed in the professional situation, engineering students seem to face challenges to make significant use of their knowledge despite rigorous exposure to and training in core competency, communication skills, multidisciplinary team work, etc. The paper aims to study the concept of self-efficacy and trace its role in shaping the engineering skills of students. The study is descriptive in nature. The findings indicate that self-efficacy has an essential role in reinforcing the learning output of engineering education.

Keywords Self-efficacy · Learning output · Engineering education

23.1 Introduction

Engineering education seems to address essential life and career skills to deal with financial pressure, career opportunities, time management, workplace etiquette, and other necessary skills. Engineering education equips the students with engineering knowledge, engineering skills, and engineering habits of mind (Costa and Kallick 2013). Engineers "make things that work" or make things "work better" (ABET 2013) using a set of multiple skills like logical thinking, problem solving, communication

A. Venugopal · R. Singh · Devika (✉)
Department of Humanities and Social Sciences, Birla Institute of Technology and Science, Pilani-Campus, Pilani 333031, RJ, India
e-mail: devika@pilani.bits-pilani.ac.in

A. Venugopal
e-mail: abhijith.venugopal@pilani.bits-pilani.ac.in

R. Singh
e-mail: rajni.singh@pilani.bits-pilani.ac.in

skills, spatial ability, technical skills, tinkering skills, and engineer design skills. Students with high engineering self-efficacy strive to acquire knowledge through task based activities, develop new skills, and apply technical academic knowledge in engineering. Students' beliefs about their capabilities in engineering-related areas influence their career choice (Lent and Brown 2006; Zeldin et al. 2008), academic behaviors and choices, goal orientation, task value, and interest in and usefulness of their tasks (Schunk et al. 2013; Bong 2001). Consequently, the demand for efficient, autonomous and competent future engineers (Holvikivi 2007) is high. These curious, optimistic, resourceful, resilient and reflective lifelong learners need to be fully prepared beyond the academic scope. Thus, ABET (Accreditation Board for Engineering and Technology) proposes training in ethics, lifelong learning, communication, and working in multidisciplinary teams to help them overcome the professional challenges. The knowledge and skills that future engineers must wield for professional execution are one of the crucial areas of research in engineering education along with technical knowledge, interpersonal and social skills.

The role of self-efficacy is widely discussed by the researchers of education and learning like Albert Bandura, Frank Pajares, Dale Schunk, Megan Tschannen-Moran, etc. Self-efficacy is "not the skills one has, but the conviction of what one can accomplish with whatever skills one possesses". It is a set of context specific beliefs about competence to perform given academic tasks at designated levels (Bandura 1997; Schunk 1985). Self-efficacy arises from the gradual acquisition of complex cognitive, social, linguistic, and/or physical skills through experience. Commonly understood as domain-specific (Schwarzer and Hallum 2008), skills-specific, task-specific and context-specific rather than a global judgment of ability, self-efficacy influences the choice, level, amount of effort and perseverance in the task performance (Bandura 1997). Academic self-efficacy affects academic achievement, academic persistence, career choices, mastery goal orientation, task value, etc. Thus, the present study aims to analyze the role of self-efficacy in engineering education by focusing on the learning outcomes of engineering education.

This study is based only on the research papers obtained from online data bases. Other sources like books, research reports and monographs are not included in this study. The study is limited to the premises of engineering skills and other technical aspects are out of the domain of this paper. Revealing the role of self-efficacy in engineering education, this study helps the engineering students/learners to practice accordingly as it influences the performance output.

Objectives of this paper are to trace the sources, influencing factors, types and instruments of self-efficacy; and to study how self-efficacy correlates with the learning output in engineering education.

23.2 Methodology

This qualitative study has a descriptive research design. The engineering education-specific research articles are obtained from secondary sources like Scopus indexed/approved online data bases. Selection criterion used specific key words related to the study like 'engineering education', 'self-efficacy', and 'engineering learning output'.

23.3 Self-efficacy: Levels, Sources, Types and Instruments

General self-efficacy is a sense of personal competence to deal with stressful situations (Schwarzer and Hallum 2008). Self-efficacy is more of a criterion-referenced evaluation of self-concept, self-belief, self-regulation and motivation (Bandura 1997; Bong and Clark 1999; Zimmerman 2000). It measures performance capabilities, not physical or psychological characteristics. It is a strong predictor of not only interest in engineering (Britner and Pajares 2006; Mau 2003) but also emotional reactions, persistence, etc. (Bandura 1997). Self-efficacy has two components: efficacy expectation (the potential to perform a particular task) and outcome expectancy (successful performance assurance). Self-efficacy has three dimensions: magnitude (level of task difficulty); strength (strong/weak conviction about magnitude); and generality (degree of generalization). The magnitude and strength of self-efficacy can influence choice of settings and activities, skill acquisition, effort expenditure, initiation, etc.

Self-efficacy becomes effective when all these dimensions and sources start influencing concurrently which also determines the level of self-efficacy.

23.3.1 High and Low Level

The two levels of self-efficacy, higher and lower, affect the performance output. Individuals with high efficacious beliefs willingly engage in challenging tasks, invest greater effort and persistence, and show superior performance output (Bong 2001; Bandura 1997). They better deal with the challenges, develop resolution strategies, exhibit higher achievement and find their work important, useful and interesting. The failure to utilize their capabilities is an indicator of low self-efficacy.

The level of self-efficacy is also dependent on the effectiveness of its sources, and magnitude, strength and generality of self-efficacy.

23.3.2 Sources of Self-efficacy

Mastery experiences mark the knowledge acquired through re-collections, accomplishments, hands-on-work experiences, and judgments on the competence of learner's previous attainments in a related task. Vicarious experiences enhance self-efficacy when engineering students see others with same potential performing the desired skills, solve the problems collectively, assess the skills of their peers, and induce a belief in their own skills to perform the same task. Verbal persuasion is encouraging judgments, constructive feedbacks and positive appraisals from teachers/role models, friends, etc. Physiological/emotive arousal is the experience of anxiety, stress, fatigue, etc. Favorable reinforcement helps positively (Bandura 1997; Ponton et al. 2001).

These sources, either individually or in combination, can affect the level of the learner's self-efficacy.

23.3.3 Types of Self-efficacy

Self-efficacy, during 1980s and 1990s, was branched into, 'career/occupational self-efficacy, (related to career aspects of self-efficacy construct like interest, career choice, etc.) (Betz and Hackett 1981), 'academic milestones self-efficacy' (a person's ability to cope through barrier situations) (Lent et al. 1986) and 'mathematics and science self-efficacy' (significantly predicts a person's science grade (Britner and Pajares 2006). Another classification of self-efficacy is: self-efficacy for self-regulated learning (self-efficacy that promotes/enhances self-learning), self-efficacy for academic achievement, course-specific self-efficacy, content specific self-efficacy and problem-specific self-efficacy (Bong 2001). Job Related Self-efficacy (self-beliefs on the capable skills which is related to different professions of engineering) comprises math skills, problem solving skills, technical/computer skills, analytical skills, and creativity. Interpersonal self-efficacy (skills that promotes different personal skills) comprises communication, teamwork, and leadership skills. Life skills self-efficacy (Life skills are those skills which would be useful in everyday life.) comprises of time management, organizational skills, and attention to details (Dent et al. 2018).

Since there are multiple types of self-efficacy, the divisions and measurements of self-efficacy also get framed accordingly.

23.3.4 Instruments of Engineering Self-efficacy

Academic Milestones measure was modified as per engineering education by including "engineering major" in the items to help students situate their self-efficacy ratings (Lent et al. 1986). General academic self-efficacy scale measures engineering students' beliefs in their capabilities to perform academically (Dunlap 2005). Self-efficacy for self-regulated learning in and academic achievement measures engineering students' performance (academic and learning) (Yildirim et al. 2010). General Engineering Self-efficacy Scale measures general engineering skills (Mamaril 2016). Engineering design self-efficacy measures specific set of design skills (Carberry et al. 2010). The Engineering Skills Assessment scale measures 11 skill areas that students believe are important for career: communication, team work, math and science, creative ability, problem solving skills, leadership and management skills, technical skills and knowledge, time management skills, analytical skills, orderliness and organizational skills, attention to details (Yildirim et al. 2010). Of all the scale mentioned above, it has been found that except Engineering Skills Assessment scale all other do not measure the engineering skills directly which reflects the need for better and more focused scales for measuring engineering skills.

Self-efficacy, with its effect on skill development, academic performance, outcome expectation, etc., contributes to the learning outcomes of engineering education. Importance of self-efficacy in skill development need not be questioned.

23.4 Learning Output of Engineering Education

The accrediting agencies, ABET and others, and researchers have stated clearly the learning outcomes of engineering education. The learning outcomes are basically dependent on skills required by engineers at their workplace. The learning outcomes discussed below, based on American Association of Engineering Societies (AAES) (Engineering Competency Model—DRAFT 2015), are divided into four areas in a broader way and further discussed on the basis of other research.

23.4.1 Personal Effectiveness

Lifelong learning. Lifelong learning/life-wide learning results from the learner's recognition of the need for and an ability to expand means of knowledge beyond the scope of formal education. An engineering context and dynamic nature of learning skills allow individuals to acquire knowledge in the best possible way (ABET 2013; Darabi 2017; Yellamraju et al. 2017).

Personal skills. It includes self-management, emotions, work-life balance/health, honesty, confidence, initiative, dependability, and reliability. Practical thinking, and binding values and attitudes towards their own ability and those of their peers (Yellamraju et al. 2017; Male et al. 2009; Baker et al. 2008) keep them self-motivated and positive.

Professionalism. Knowing about recent trends of the field, networking, meeting deadlines, adapting to change, willing and persistent to try new processes, they exhibit leadership qualities. Successful conflict resolution and action orientation help engineers take decisions with time and knowledge constrains, avoid delays, and maintain a sense of urgency (ABET 2013; Male et al. 2009; Direito et al. 2012).

Personal effectiveness deals with the interpersonal self-efficacy which promotes interpersonal skills and life skills. Learners can use their self-regulated learning to regulate their personal effectiveness.

23.4.2 Industry-Wide Technical

Foundations of Engineering. Understanding the relation between theory and application (Baker et al. 2008), they use the acquired knowledge to understand the impact of engineering solutions in a global, economic, environmental, and societal context (Male et al. 2009; Carlson 2018; Loveland and Dunn 2014; González-Lizardo 2008).

Design. Design processes support active and self-directed learning; learners imagine and understand how a part works together in the first view itself (ABET 2013; Carberry et al. 2010; Baker et al. 2008; Lucas et al. 2014).

Sustainability and Social and Environmental Impact. Engineering education provides the learning ground to formulate, analyze, advocate health and safety issues, take considered risks. Responsibilities include awareness of life-cycle principles, environmental impact, sustainability, and professional practices (ABET 2013; Male et al. 2009; Carlson 2018).

Professional Ethics. Impact of any new product or design (unexpected and undesirable) on people, systems, and the environment must be carefully sensitised by a prism of myriad possibilities. Ethically conscious, they conduct risk analysis on their research designs – for example, considering disposal of waste from the engineering process. (ABET 2013; Carlson 2018; Loveland and Dunn 2014; Lucas et al. 2014; Grubbs et al. 2018).

Industry wide technical is the combination of occupational/career self-efficacy, academic milestone self-efficacy, interpersonal self-efficacy and job skills self-efficacy. It provides a broad and inclusive understanding in different areas where execution of tasks poses real, different and unpracticed challenges.

23.4.3 Workplace

Team Work. It demands managing projects and contracts in diverse discipline and fields using strategic planning, mentoring and interacting with people, showing concern for others, voluntarily sharing information, ensuring fare decisions and facilitating their individual contribution (ABET 2013; Dent et al. 2018; Male et al. 2009; Loveland and Dunn 2014; NAE Annual 2010).

Working with Tools and Technology. The most important learning is to distinguish between unquestioned acceptance for technological change and thoughtful understanding of all aspects of change (ABET 2013; Baker et al. 2008; Loveland and Dunn 2014).

Problem Solving. Problem solving is not same as problem finding. Listening to and recitation of details are essential for sophisticated problem solving (Lucas et al. 2014; Nordstrom and Korpelainen 2011). Open-ended problem solving is an efficacious tool of practical thinking (Loveland and Dunn 2014). Formulating and analysing the model of a problem, doing back-of-the-envelope calculations to show workability (Baker et al. 2008), and asserting approaches or needs develop a path to solution (Male et al. 2009).

Creative Thinking. Flexibility, originality, fluency, and associative thinking mark thinking about something in a novel, diverse, quick and different way. Using elaboration, brainstorming, modification, and attribute listing, learners explore, engage, and evaluate to generate, present and sum up convincingly (Baker et al. 2008; Loveland and Dunn 2014).

Different learning outputs under workplace are attributed to multiple aspects such as individual and collective effort making, knowledge of subject, instruments, and challenging situations. Workplace learning outputs directly contribute to the job-related self-efficacy.

23.4.4 Academic

Communication. Communication skills (oral and written) comprise background explanation and meaningful flow of ideas (Yellamraju et al. 2017; Grubbs et al. 2018). Communication, a process in which people try to inform and educate is an essentiality of engineering habits of mind. Professionals need to remain up-to-date (Male et al. 2009).

Mathematics. There is a need of Mathematics self-efficacy to handle mathematical rigor, remember details of definition, understand and apply mathematical concepts to problems. Deductive reasoning skills are used to select an appropriate computation and estimation method to carry out the mathematical processes accurately (Britner and Pajares 2006; Yellamraju et al. 2017; Baker et al. 2008).

Sciences. As a bridge between conceptual and practical learning, basic sciences contribute to creative/practical/utilitarian engineering practices. Systematic content knowledge (multiple angles, breadth and depth of knowledge) is used to overcome multiple realistic challenges (ABET 2013; Zeldin et al. 2008; Loveland and Dunn 2014; Lucas et al. 2014).

Critical and Analytical Thinking. Engineering students use cause-effect, patterning, analogies, forecasting, critiquing, etc. to detect the symptoms of doubtful solutions, assertions and arguments at work (Yellamraju et al. 2017; Loveland and Dunn 2014). The practical experiments make them handle different domain specific situations as practicing professionals (Baker et al. 2008; Loveland and Dunn 2014).

In academic, knowledge of mathematics and sciences prepares the foundation of engineering education. Engineering education foundation is also supported by communication skills and critical thinking.

All the four areas of learning output are directly related to different sources and types of self-efficacy. The analysis of self-efficacy and learning outcomes of engineering education highlights the relationship between self-efficacy and learning outcomes. The different types of self-efficacy have different roles to play in different skills developed during engineering education.

23.5 Conclusions

Self-efficacy significantly affects the learning output of engineering education. During the process of learning, learners, using self-efficacy construct their own knowledge of disciplinary field, evolve and/or modify the existing knowledge contextually. The conviction of the learners, the active agents of learning, about what they can accomplish with their skills enables them to gain experience using flexible, relative and additive understanding of their knowledge. Conducive learning environment can be used as an asset by the learners who wish to explore new avenues of learning/knowledge. Integrating the previously gained experiences with the recently acquired knowledge, the learner can better understand the differences and nuances of learning. Reflective thinking can contribute to self-regulated learning from contextualising and decontextualising the situations.

References

ABET (2013) Criteria for accrediting engineering programs. http://www.abet.org/wp-content/uploads/2015/04/eac-criteria-2013-2014.pdf

Baker D, Krause S, Purzer SY (2008) AC 2008-633: developing an instrument to measure tinkering and technical self-efficacy in engineering. ASEE

Bandura A (1997) Self-efficacy: the exercise of control. Freeman, New York

Betz NE, Hackett G (1981) The relationship of career-related self-efficacy expectations to perceived career options in college women and men. J Counsel Psychol 28:399–410

Bong M (2001) Role of self-efficacy and task-value in predicting college students' course performance and future enrollment intentions. Contemp Educ Psychol 26:553–570. https://doi.org/10.1006/ceps.2000.1048

Bong M, Clark RE (1999) Comparison between self-concept and self-efficacy in academic motivation research. Educ Psychol 34:139–154. https://doi.org/10.1207/s15326985ep3403_1

Britner SL, Pajares F (2006) Sources of science self-efficacy beliefs of middle school students. JRST 43:485–499

Carberry AR, Lee HS, Ohland MW (2010) Measuring engineering design self-efficacy. J Eng Educ 99:71–80

Carlson WB (2018) Knowledge, skill, and wisdom: reflections on integrating the social sciences and engineering. In: Subrahmanian E, Odumosu T, Tsao JY (eds) Engineering; A better future interplay between engineering, social sciences, and innovation. Springer, United States, pp 187–196

Costa A, Kallick B (2013) Dispositions: reframing teaching and learning. Professor John Perkins' review of engineering skills, London

Darabi H et al (2017) Life after university for engineering graduates. In: 2017 ASEE annual conference and exposition, ASEE

Dent L, Maloney P, Karp T (2018) Self-efficacy development among students enrolled in an engineering service-learning section. IJSLE, Human Eng Soc Entrepre 13:25–44. https://doi.org/10.24908/ijsle.v13i2.11483

Direito I, Pereira A, Duarte A (2012) Engineering undergraduates' perceptions of soft skills: relations with self-efficacy and learning styles. Soc Behav Sci 55. https://doi.org/10.1016/j.sbspro.2012.09.571

Dunlap JC (2005) ETR&D 53:65. https://doi.org/10.1007/BF02504858

Engineering Competency Model—DRAFT (2015) January. http://www.aaes.org/sites/default/files/Engineering_Competency_Model_draft-0115.pdf
González-Lizardo AE (2008) Interdisciplinary research for engineering skills development. Acta Universitaria Direccion De Invesyigacion Y Postgrado 18(3), University of Guanajuato
Grubbs EM, Strimel GJ, Huffman T (2018) Engineering education: a clear content base for standards. TET 33-38, ITEEA
Holvikivi J (2007) Learning styles in engineering education: the quest to improve didactic practices. Europ J Eng Educ 32(4):401–408. https://doi.org/10.1080/03043790701332909
Lent RW, Brown SD (2006) On conceptualizing and assessing social cognitive constructs in career research: a measurement guide. J Career Assess 14:12–35. https://doi.org/10.1177/1069072705281364
Lent RW, Brown SD, Larkin KC (1986) Self-efficacy in the prediction of academic performance and perceived career options. J Counsel Psychol 33:265–269
Loveland T, Dunn D (2014) Teaching engineering habits of mind in technology education. TET 73(8):13–19
Lucas B, Hanson J, Claxton G (2014) Thinking like an engineer: implications for the education system. RAE, London
Male SA, Bush MB, Chapman ES (2009) Identification of competencies required by engineers graduating in Australia. In: Kestell C, Grainger S, Cheung J (eds) 20th conference of the AAEE: engineering the curriculum. University of Adelaide, pp 882–887
Mamaril A et al (2016) Measuring undergraduate students' engineering self-efficacy: a validation study. J Eng Educ. https://doi.org/10.1002/jee.2012
Mau WC (2003) Factors that influence persistence in science and engineering career aspirations. Career Develop Quart 51(3):234–243. https://doi.org/10.1002/j.2161-0045.2003.tb00604.x
NAE Annual Report (2010). https://www.nae.edu/51959/NAE-Annual-Report-2010
Nordstrom K, Korpelainen P (2011) Creativity and inspiration for problem solving in engineering education. Teach Higher Educ 16(4):439–450. https://doi.org/10.1080/13562517.2011.560379
Ponton M, Horine EJ, Ukeiley LM, Seiner J (2001) Understanding the role of self-efficacy in engineering education. J Eng Educ. https://doi.org/10.1002/j.2168-9830.2001.tb00599.x
Schunk DH (1985) Self-efficacy and classroom learning. Psychol Sch 22:208–223. https://doi.org/10.1002/1520-6807(198504)22:2%3c208:AID-PITS2310220215%3e3.0.CO;2-7
Schunk DH, Meece JR, Pintrich PR (2013) Motivation in education: theory, research, and applications, 4th edn. Pearson, United States
Schwarzer R, Hallum S (2008) Perceived teacher self-efficacy as a predictor of job stress and burnout: mediation analyses. Appl Psychol 57:152–171. https://doi.org/10.1111/j.1464-0597.2008.00359.x
Yellamraju T, Magana AJ, Boutin M (2017) Investigating engineering students habits of mind: a case study approach. In: 2017 ASEE ACE, ASEE
Yildirim T, Sacre MB, Shuman L (2010) AC 2010-627: scale development for engineering modeling self-efficacy. ASEE
Zeldin AL, Britner SL, Pajares P (2008) A comparative study of the self-efficacy beliefs of successful men and women in mathematics, science, and technology careers. JRST 45(9):1036–1058. https://doi.org/10.1002/tea.20195
Zimmerman BJ (2000) Self-efficacy: an essential motive to learn. Contemp Educ Psychol 25:82–91. https://doi.org/10.1006/ceps.1999.1016

24

Engineering Education: Implementing Collaborative and Experiential Learning

Manoj Kumar Soni and Tamali Bhattacharya

Abstract The conventional teaching method is mostly as a one-way communication be-tween instructor and students, where students go through rote learning and then appear for written exam. Based on their performance in written exam they are graded and judged. This type of system inculcates exam oriented learning approach than to actual knowledge gain. This is a major issue in the present era where students are fed up with exam oriented approach. So there is immense need to bring innovation in teaching methods. Also, along with education, among the students, there is need of growing social awareness specially towards weaker section of society. In order to impart effective learning experiential and collaborative learning has been incorporated in solar energy related courses by the authors at BITS Pilani. During these courses, the students fabricated do-it-yourself type solar cookers and tested the performance of these cookers in normal conditions. In order to have feel of the usage of the fabricated cookers the students were asked to cook some food items using those cookers. After successful completion of these activities, with the help of a NGO students helped the authors to conduct cheap solar cooker workshop for village women at a village near the institute. During the workshop the students helped the village women to fabricate cheap solar cookers. All the activities fabrication, testing and involvement in the workshop was the part of evaluative components for the courses. It was overall a very successful experiential and collaborative learning experiment with social angle involved in it.

Keywords Collaborative learning · Experiential learning · Societal awareness

M. K. Soni (✉) · T. Bhattacharya
Birla Institute of Technology and Science, Pilani Campus, Pilani 333031, RJ, India
e-mail: mssoni@pilani.bits-pilani.ac.in

24.1 Introduction

In 21st century, skill development is necessary for the Engineering graduates. The total development of skills is very important—creative skills, cooperative skills, management skills, communication skills, leadership skills, teamwork, which develop the positive behavior and attitude to the learner. It is an important mission for the teachers to improve their students' skills in real-life situations and working for the society. Outcome-based education keeps learner at the center of the education system and teacher acts as a facilitator to help them in their learning. Learning by doing has been a principle of thousands of years (Reese 2011). The learning-by-doing principle has been supported widely and in many forms-mainly learn-by-doing, trial-and-error learning, practical experience versus book learning, the practice-theory-practice dialectic and "proof upon practice."

Dale's Cone of Experience is a model that combines several theories related to instructional design and learning processes. During the 1960s, Dale theorized that learners retain more information by what they "do" as opposed to what is "heard", "read" or "observed" (Davis and Summers 2015). At present, this "learning by doing" has become known as "experiential learning" or "action learning" (http://www.queensu.ca/teachingandlearning/modules/active/documents/Dales_Cone_of_Experience_summary.pdf).

According to Dale's research, the least effective method at the top, which encompasses learning from information presented through reading, hearing, view images, watching videos. The most active method is at the bottom, which involves direct, purposeful learning experiences, such as hands-on or field experience. Direct purposeful experiences represent reality or the closet things to real, everyday life. "Action-learning" techniques result in up to 90% retention. Here we will examine student projects in two elective subjects "Renewable Energy" and "Solar Thermal Process Engineering" course at Birla Institute of Technology and Science (BITS) Pilani, at Pilani campus in which students' evaluation was done by mid semester and end semester results, project evaluation and from the surveys. The author surveys students project evaluation and measure student achievement of learning outcomes.

The author introduces cooperative and collaborative learning as a teaching learning strategy which means the students grouped together to inspect a particular assignment or topic to maximize their own and each other's learning. Collaborative is an informal method of teaching and learning, based on the idea that learning is a natural social act in which the participants talk, share, plan together to fulfil the learning goal. Collaborative principles are based on the theories of Dewey (2009), Vygotsky (1980), and Bloom (1956). Their collective work focusing on how students learn has led educators to develop more student-focused learning environments that put students at the center of instruction. The learner is the primary focus in collaborative activities, i.e. working in groups, where the main objective is learning by doing. The structured approaches to developing solutions to the real world problems incorporated in the learning (Theodore 1999; Dewey 2009; Vygotsky 1980; Bloom 1956).

The authors focus to project based learning, which is a form of active learning, based on inter disciplinary means combination of two to more academic disciplines into an activity. This type of pedagogy mainly applies to learner' life and teacher considers to learner' need and interests. Learners need to be educated in such a way that they work for their community and be responsible for community affairs.

Experiential learning helps in learning course quickly as it is learning while doing. This was tried in solar energy related course. In this total of about 40 students were involved. The basic concepts about solar energy and solar cookers are discussed in the class.

24.2 Fabrication of Solar Cooker

The students were divided into small teams comprising of minimum two to maximum four students. The team formation was left to the students as per their choice. They informed the instructor once their teams are formed. The do-it-yourself designs of various solar cookers are shared with the students. They gone through various design and selected suitable design for fabrication. Students were asked to first fabricate a small size model using paper or cardboard to understand intricacies involved in the actual fabrication purpose. For fabrication of prototype the corrugated plastic sheets, anodized aluminum reflector sheets and required tools were made available to the students in the institute workshop. The students went to workshop as during working hours as per their free time between classes and fabricated various solar cookers using corrugated plastic sheets for strength and then mounted anodized aluminum sheets on the sheets. Figure 24.1 represents the pictures of the fabrication of solar cookers by the students. During fabrication, students have to manage their schedule in such a fashion that they are attending classes as well as during office hours and their free time they had to manage the entire fabrication process.

24.3 Testing of Solar Cookers

After fabrication, the solar cookers were tested by students. For the same water, cooking pot, thermocouples and data logger were made available to the students. The students used transparent polythene bag for creating greenhouse effect. The radiation data was recorded using Pyranometer. The students recorded data for certain time period and then evaluated the performance of each solar cooker using recorded data. The testing of solar cookers is represented by pictures in Fig. 24.2.

Fig. 24.1 Fabrication of solar cookers by the students

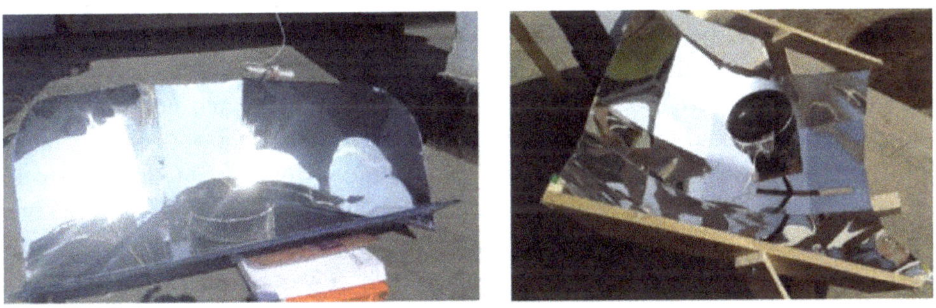

Fig. 24.2 Testing of solar cookers

24.4 Cooking Using Solar Cookers

Whenever a product is made it is important to understand its usability and related aspects from consumer's point of view. Keeping this in mind, the students were asked to now use the solar cooker they fabricated by cooking some simple food like noodles, rice, etc. as they are not expert in cooking. They were very excited and cooked noodles and rice using solar cookers they fabricated. The students also noted down the time required to cook the food. Students were asked to cook with polythene

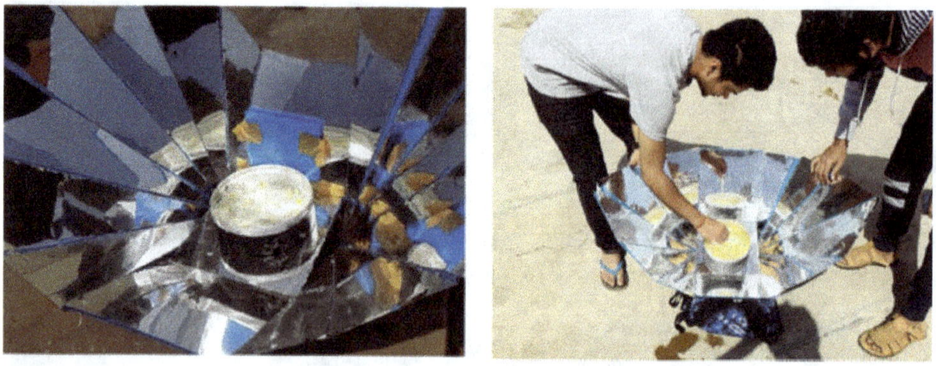

Fig. 24.3 Cooking using solar cookers

cover and without cover both in order to understand the importance of minimizing losses (Fig. 24.3).

24.5 Workshop for Village Women

Apart from involved in quality research and imparting technical education, educational institutes should inculcate social responsibility among the student. BITS Pilani is located in a small town Pilani in Rajasthan. It is surrounded by many small villages. In villages, cooking is mainly done on chullah by burning wood. This type of coking has harmful impact on the health of women due to the exposure to fumes and exhaust gases. In order to instill social responsibility in the students and help the village women to reduce cooking with chullah and associate problems, a workshop on fabrication of cheap solar cookers (@ 1$) was conducted for them with the help of students registered in the course and Nirmaan Organization. Nirmaan is a NGO that was started by BITS Pilani students in 2005. It mainly works in the areas of education, livelihoods and social leadership. The workshop was conducted on 1st April, 2018 at Khedla ka Baas village which is three kilometers from the campus. About 33 village women and 15 kids were present in the workshop.

For fabrication of solar cookers cardboard boxes and cheap Reflective Mylar sheet were used. The workshop was conducted with an aim to educate village women about usage of solar energy using solar cooker and show its effectiveness and give the women an alternative to fuel wood for cooking. All the students registered in the course participated enthusiastically as this was unique activity apart from regular classroom learning. In the workshop, they taught the women how to make reflective solar cookers. During the workshop, students and women were divided into groups with kids of the village also joining in. The groups were made in such a fashion that every group has at least two to three students along with village women. The students made solar cookers in front of the women and then gave the women the chance to make the cookers by themselves. Majority of the women as well as young

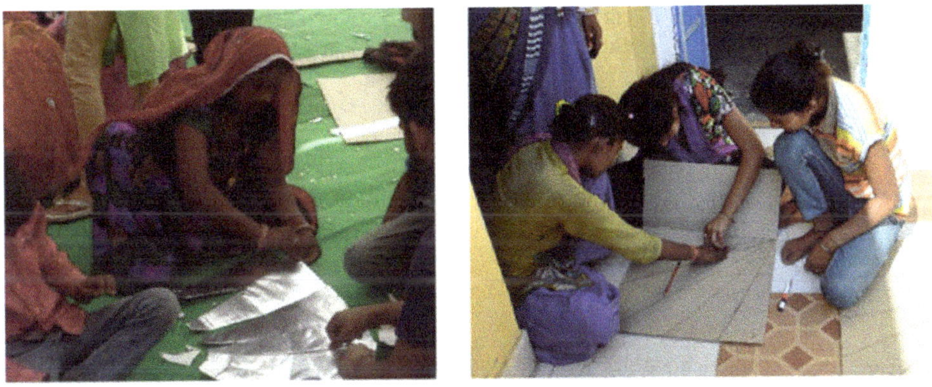

Fig. 24.4 Workshop of village women

kids learned to make these reflective solar cookers by themselves. A demonstration of one of the pre-fabricated solar cookers was given by heating water and later making Khichdi with that cooker. The participants appreciated the technology and realized its benefits over traditional wood burning as it will help them to have healthier cooking environment. This will improve their health as they avoid fumes etc. coming out of burning wood. Few volunteers from Nirmaan also helped in conducting the workshop smoothly. After completion of the workshop all the women were given solar cookers that they had made by themselves for future use. This entire activity was one of the evaluative component of the course. Apart from the evaluation component the students realized their social responsibilities and also learned about the traditional conditions in which village women cook the food. Students also expressed their willingness to participate in future workshops. Figure 24.4. Gives some glimpses of the workshop conducted for the village women with students' involvement.

24.6 Feedback Survey

In order to assess the impact of all these activities related to the course, an online feedback survey was conducted. For this, a google form consisting of eight statements was designed and the link was shared with the students through email. The students were asked to give their feedback on 5 point Likert scale with 1 being "strongly disagree", 2 being "disagree", 3 being "Neither agree nor disagree", 4 being "agree" and 5 being "strongly agree". The feedback was collected and the result of the feedback survey given by the students is as shown in Fig. 24.5a–h in the form of bar chart. The feedback survey indicates that majority of students consider this way of learning as very effective, it helped them to work independently, getting involved in multidisciplinary and social activity. This method has also helped them in engaging actively in the class room as it is effective and useful tool to learn complex scientific knowledge easily as compared to the traditional ways. Also due to it their interest towards the course has increased which helped in improvement in their learning.

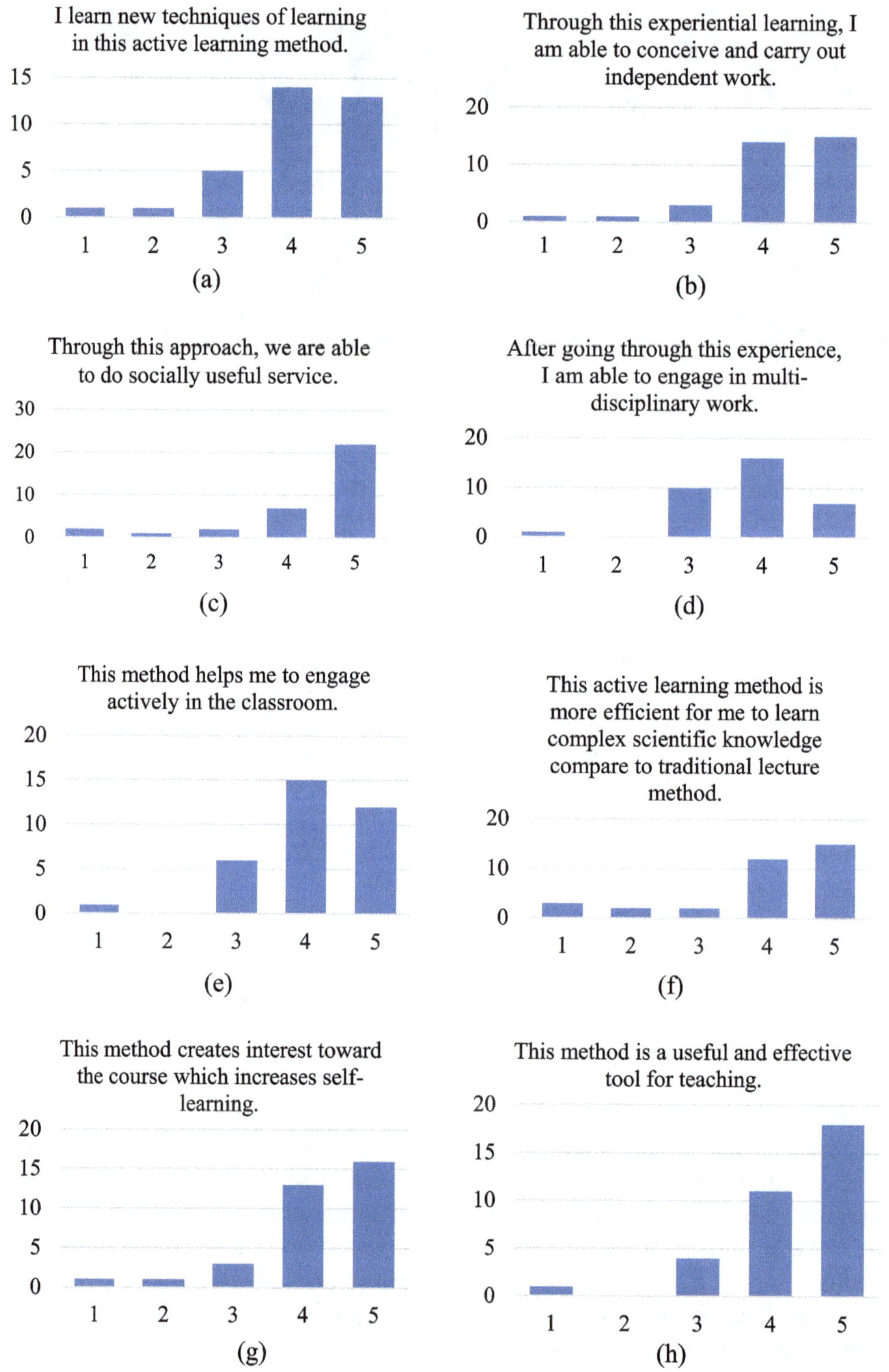

Fig. 24.5 The feedback survey

24.7 Conclusions

There is immense need to bring innovation in teaching methods for the present generation. Also, among the students, there is need of growing social awareness specially towards weaker section of society as enhancement of the society is very important part in the economy. For effective learning, experiential and collaborative learning has been incorporated in solar energy related courses by the authors at BITS Pilani. Students fabricated do-it-yourself type solar cookers, tested the performance of these cookers in normal conditions and cooked food items using those cookers. With the help of a NGO, students helped the authors to conduct cheap solar cooker workshop for village women at a village near the institute. During the workshop, the students helped the village women to fabricate cheap solar cookers. All the activities fabrication, testing and involvement in the workshop was the part of evaluative components for the courses. It was overall a very successful experiential and collaborative learning experiment along with societal awareness and participation.

References

Bloom BS (Ed) (1956) Taxonomy of educational objectives. Handbook 1: cognitive domain. WhitePlains, NY, Longman. https://www.uky.edu/~rsand1/china2018/texts/Bloom%20et%20al%20-Taxonomy%20of%20Educational%20Objectives.pdf. Last accessed 3 July 2019

Davis B, Summers M (2015) Applying Dale's cone of experience to increase learning and retention: a study of student learning in a foundational leadership course. QSci Proc 2015:6. https://doi.org/10.5339/qproc.2015.wcee2014.6

Dewey J (2009) Democracy and education: an introduction to the philosophy of education. https://s3.amazonaws.com/arena-attachments/190319/2a5836b93124f200790476e08ecc4232.pdf. Last accessed 3 July 2019

Heidi MA, Dale's cone of experience. http://www.queensu.ca/teachingandlearning/modules/active/documents/Dales_Cone_of_Experience_summary.pdf. Last accessed 3 July 2019

Reese HW (2011) The learning-by-doing principle. Behav Develop Bull 17(1):1–19. https://doi.org/10.1037/h0100597

Theodore P (1999) Collaborative versus cooperative learning: a comparison of the two concepts which will help us understand the underlying nature of interactive learning. ERIC Institute of Education Science, pp 1–13. https://files.eric.ed.gov/fulltext/ED448443.pdf. Last accessed 3 July 2019

Vygotsky L (1980) Mind in society: the development of higher psychological processes. Harvard University Press, Cambridge, MA. http://ouleft.org/wp-content/uploads/Vygotsky-Mind-in-Society.pdf. Last accessed 3 July 2019

25

CFD Methods: Application of Lateral Thinking Techniques

Priya C. Sande

Abstract Although a traditional engineering education system is good in imparting the intended know-how, it does not focus much on the delivery system. Applying lateral thinking techniques can provide a value addition to delivery as illustrated in this work, for the course of Computational Fluid Dynamics (CFD). CFD is a powerful engineering simulation tool and a trending elective course taught at several universities. CFD methods build on numerical methods to resolve flows and design flow related equipment. Here several methods form the core of the course, making teaching and learning them all a rigorous mathematical exercise. Hence lateral thinking techniques were applied to sustain and enhance student interest. A basic CFD method (Lax-Wendroff) was first taught. It was then illustrated how *suspended judgment*, *generation of alternatives* and *fractionation* were applied to further elucidate closely related CFD methods, namely those of Maccormack and Richtmyre. Hence lateral thinking was successfully applied for value addition to delivery in the class room, and a framework was thereby suggested to explore other CFD methods in this way. The work demonstrates the possibilities in amalgamating lateral thinking with analytical content in engineering education.

Keywords Lateral thinking techniques · CFD methods · Creative engineering pedagogy

25.1 Introduction

Traditionally, it is assumed that creativity is more associated, understood and more frequently spoken about in the Arts (Heilmann and Korte 2010), thus the term 'creative arts'. But William J. J. Gordon affirms that the intellectual processes that characterize creativity are the same whether applied in arts, science or even engineering (Gordon 1961). Edward de Bono the pioneer in creativity, propagated the

P. C. Sande (✉)
Department of Chemical Engineering, Birla Institute of Technology and Science, Pilani-Campus, RJ 333031 Pilani, India
e-mail: priya@pilani.bits-pilani.ac.in

term 'lateral thinking' instead of 'creative thinking' (De Bono 1970), and these terms will hence be used interchangeably in this work. His work describes the advantages and means of applying lateral thinking to the process of design, which is at the very heart of engineering. He affirms that while vertical thinking is excellent, it is not sufficient. Lateral thinking provides enhanced value. Values can be equated with benefits, which when propagated can arguably 'complete' the process of knowledge dissemination which we have come to call 'education'. If vertical thinking is concerned with selecting and using standard patters (or solutions) for problem solving, then lateral thinking is all about reconstructing and transforming information into new patterns. Even though the importance of creative thinking in academic settings has been stressed for a long time (Senra and Fogler 2014), pedagogy in engineering courses has largely revolved around vertical thinking (Tekic et al. 2015). This is not surprising, considering the analytical nature of the courses. With little agreement on the definition of creativity or how to teach it, firsthand information about creativity and the creative process, as experienced by working engineers was studied (Klukken et al. 1997). Creativity in a general sense has been researched in engineering education, with one study concluding that creativity in engineering curricula is not appropriately taught or rewarded, and that students who view themselves as highly creative were actually less likely to graduate in engineering (Atwood and Pretz 2016). Another work which aimed to document opportunities for creative growth, found that convergent thinking was well represented in the engineering courses, but generation of ideas, divergent thinking and idea exploration was largely lacking (Daly et al. 2014). However in one case study on mechanical engineering project work, special attention was paid to the nature of creativity, with exercises introduced to facilitate this historically neglected aspect of engineering education (Conwell et al. 1993).

Whereas creativity is difficult to define and too often only the description of a result, lateral thinking is the description of a process (De Bono 1970) and described in the next section. In this work the lateral thinking techniques proposed by de Bono were adapted for the first time in pedagogy for the course 'Computational Fluid Dynamics' (CFD). The work is a novel attempts at integrating creative thinking with the mathematical methods of CFD course, thereby opening up new possibilities for extending such pedagogy in engineering education. The benefit to the student is two fold in terms of easier and stronger grasp of the complex and varied CFD methods, and familiarity with lateral thinking techniques, which have broad application in problem solving.

25.2 Some Lateral Thinking Techniques

Lateral thinking as the name suggests, can be best viewed as complementary or even opposite in some sense to vertical thinking. The action of vertical thinking is mainly related to selection, rejection and development of arranged bits of information called patterns, whereas lateral thinking acts to restructure the information bits into completely different patterns. Lateral thinking involves escape from the old, and provocation of new patterns, so it is 'provocative' while vertical thinking is 'analytical'. With regard to the way the mind works: vertical thinking results from familiar pattern reinforcement and stocking. These patterns become difficult to change once established, so a deliberate application of lateral thinking techniques aims to restructure patterns and put information together in new ways. The following three techniques of de Bono's several lateral thinking techniques (De Bono 1970) were applied in this work.

25.2.1 Suspended Judgment

This is a lateral thinking effort, in which one does not try to size-up an idea quickly, but rather lets it live on with the intention of converting it into a better idea. This runs contrary to vertical thinking, where ideas which are deemed unwise, impractical, fanciful or fantastic are snuffed out at their inception and are not allowed to advance in the design process. This is because correct ideas are valued in vertical thinking over a diversity of ideas, as in creative thinking.

25.2.2 Generation of Alternatives

Of all techniques, the generation of alternatives is the one which forms the core of creative thinking. Here there is a deliberate search and exploration of alternatives, which goes beyond a mere natural inclination which usually halts when a promising approach presents itself. Instead a wide ranging search is engaged in, to gather a diversity of ideas. Outcomes could be: (a) one of the generated ideas leads to the best idea; (b) the original idea is still chosen, but for the reason that it is indeed the best among many; (c) generation of ideas may simply loosen up rigid patterns, open-up new information or challenge assumptions which provoke further idea generation.

25.2.3 Fractionation

Fractionation enables new views of a situation for better and easier generation of solutions. Usually the commonly used standard patterns were originally developed from smaller patters that are no longer referred to. To restructure and bring about insight one may have to go back and break-up patterns into fractions, which can be further restructured into new options or solutions.

25.3 Illustration of CFD Methods Using Lateral Thinking

CFD is an approach to solving flow problems by numerically resolving the governing (Navier-Stokes) partial differential equations, over discrete grid points, rather than analytically over a continuum. In an introductory CFD course, all aspects of the governing equations and grid generation are dealt with in the first half, followed by the all important computational methods. Although commercial CFD software platforms use variants of the Pressure Correction method, understanding of several precursor methods is paramount; not only to understand evolution of the methods, but also to grasp mature methods like Pressure Correction. The Lax-Wendroff (Lax-W) method is conceptually the simplest (only two main steps) and probably the easiest for a novice to grasp. It is also usually taught as the first of many methods in a CFD course, and several methods evolved as improvements. Teaching and learning all these methods can be laborious due to their complexities and similarities. By introducing lateral thinking, it is intended not only that the learning process would thereby become less arduous and more interesting, but also that lateral thinking techniques would be conveyed as an added benefit. To illustrate, two improvement methods are detailed as derived from the fundamental Lax-W method, using the said lateral thinking techniques. Standard CFD textbook, Anderson (Anderson 2012) covers these methods in detail.

25.3.1 Maccormack Method Illustrated from Lax-Wendroff Method

The Lax-W method, is conceptually the simplest method, in the case that flow variable (density (ρ), *x*-component of velocity (u), *y*-component of velocity (v) and specific internal energy (e)) are know at initial time *t*. In this method the standard continuity, *x*-momentum, *y*-momentum and energy equations for compressible fluid are modified and used in the form of Eqs. (25.1), (25.2), (25.3), and (25.4) respectively:

$$\frac{\partial \rho}{\partial t} = -\left(\rho\frac{\partial u}{\partial x} + u\frac{\partial \rho}{\partial x} + \rho\frac{\partial v}{\partial y} + v\frac{\partial \rho}{\partial y}\right) \tag{25.1}$$

$$\frac{\partial u}{\partial t} = -\left(u\frac{\partial u}{\partial x} + v\frac{\partial u}{\partial y} + \frac{1}{\rho}\frac{\partial p}{\partial x}\right) \tag{25.2}$$

$$\frac{\partial v}{\partial t} = -\left(u\frac{\partial v}{\partial x} + v\frac{\partial v}{\partial y} + \frac{1}{\rho}\frac{\partial p}{\partial y}\right) \tag{25.3}$$

$$\frac{\partial e}{\partial t} = -\left(u\frac{\partial e}{\partial x} + v\frac{\partial e}{\partial y} + \frac{p}{\rho}\frac{\partial u}{\partial x} + \frac{p}{\rho}\frac{\partial v}{\partial y}\right) \tag{25.4}$$

To derive the flow variables (for example for a ρ) at the next time step '$t + \Delta t$' (called marching solution), and for all grid points (i, j), the Taylor series expansion including the second derivative (to maintain second-order solution accuracy) is used:

$$\rho_{i,j}^{t+\Delta t} = \rho_{i,j}^{t} + \left(\frac{\partial \rho}{\partial t}\right)_{i,j}^{t} \Delta t + \left(\frac{\partial^2 \rho}{\partial t^2}\right)_{i,j}^{t} \frac{(\Delta t)^2}{2} + \cdots \tag{25.5}$$

Similar equations are written for all flow variables. The first derivative is substituted directly from governing Eqs. (25.1)–(25.4), and a second application of differentiation operator on the same equations provides the second derivatives, as for example Eq. (25.6).

$$\begin{aligned}\frac{\partial^2 \rho}{\partial t^2} = -&\rho\frac{\partial^2 u}{\partial x \partial t} + \frac{\partial u}{\partial x}\frac{\partial \rho}{\partial t} + u\frac{\partial^2 \rho}{\partial x \partial t} + \frac{\partial \rho}{\partial x}\frac{\partial u}{\partial t} \\ +&\rho\frac{\partial^2 v}{\partial y \partial t} + \frac{\partial v}{\partial y}\frac{\partial \rho}{\partial t} + v\frac{\partial^2 \rho}{\partial y \partial t} + \frac{\partial \rho}{\partial y}\frac{\partial v}{\partial t}\end{aligned} \tag{25.6}$$

The mixed derivatives on the right-hand-side of Eq. (25.6) can similarly be derived from governing equations. Subsequently, the determined first and second derivatives are written using their second-order central difference forms and substituted in Eq. (25.5) (not shown for brevity) to provide the required solution at next time step.

Hence the essence of Lax-W method is that: with only current time data inputs, a mathematical device (Taylor series expansion) advances the solution to the next time step, with the physics injected in by the time derivates, which are appropriately derived and substituted from governing equations with spatial derivatives.

The Maccormack (Mac) method can be viewed and easily remembered as a lateral thinking variant of Lax-W method. First, an effort at *suspended judgment* offers a simplified Taylor series expansion that includes only the first derivative, as in Eq. (25.7). Such an approximation assumes straight line relationship of a variable between time steps, which would otherwise be rejected because it offers only first-order solution accuracy. On the other hand omitting the second derivative drastically cuts computational. From here it can easily be seen that a *better* time derivative (designated with subscript 'av' in Eq. (25.8)) would enhance utility of Eq. (25.7); so thinking to *generate alternative* derivatives unfolds as an aim, which leads to the possibility of constructing this '*better* time derivative' from '*better* predicted variables' (designated with bar on top, as given by Eq. (25.9) for example).

$$\rho_{i,j}^{t+\Delta t} = \rho_{i,j}^{t} + \left(\frac{\partial \rho}{\partial t}\right)\Delta t \tag{25.7}$$

$$\rho_{i,j}^{t+\Delta t} = \rho_{i,j}^{t} + \left(\frac{\partial \rho}{\partial t}\right)_{av}\Delta \tag{25.8}$$

$$(\bar{\rho})_{i,j}^{t+\Delta t} = \rho_{i,j}^{t} + \left(\frac{\partial \rho}{\partial t}\right)_{i,j}^{t} \Delta t \tag{25.9}$$

When these predicted variables for next time step, obtained from Taylor series simplification (only first two terms), are substituted for example in the difference equation form of continuity equation, a 'better alternative time derivative' for ρ is obtained:

$$\left(\frac{\overline{\partial \rho}}{\partial t}\right)_{i,j}^{t+\Delta t} = -\left[(\bar{\rho})_{i,j}^{t+\Delta t}\frac{(\bar{u})_{i,j}^{t+\Delta t} - (\bar{u})_{i-1,j}^{t+\Delta t}}{\Delta x} + (\bar{u})_{i,j}^{t+\Delta t}\frac{(\bar{\rho})_{i,j}^{t+\Delta t} - (\bar{\rho})_{i-1,j}^{t+\Delta t}}{\Delta x}\right.$$
$$\left. +(\bar{\rho})_{i,j}^{t+\Delta t}\frac{(\bar{v})_{i,j}^{t+\Delta t} - (\bar{v})_{i,j-1}^{t+\Delta t}}{\Delta y} + (\bar{v})_{i,j}^{t+\Delta t}\frac{(\bar{\rho})_{i,j}^{t+\Delta t} - (\bar{\rho})_{i,j-1}^{t+\Delta t}}{\Delta y}\right] \tag{25.10}$$

Since average of the derivative at t (using forward differencing) and derivative using predicted variables at $t + \Delta t$ (using backward differencing) is taken (Eq. (25.11)), information from both grid points lying on the two sides of grid point i, j is included in the same way as in the central differencing employed in Lax-W method. Hence both methods are second order accurate.

$$\left(\frac{\partial \rho}{\partial t}\right)_{av} = \frac{1}{2}\left[\left(\frac{\partial \rho}{\partial t}\right)_{i,j}^{t} + \left(\frac{\overline{\partial \rho}}{\partial t}\right)_{i,j}^{t+\Delta t}\right] \tag{25.11}$$

25.3.2 *Richtmyer Method Illustrated from Lax-Wendroff Method*

The Richtmyer method, also called two-step Lax-Wendroff method, was proposed for the case of non-linear equations where the substitution of temporal with spatial derivatives no longer remains straightforward or unique. This method can also be viewed and easily remembered as a lateral thinking variant of Lax-W method. First, the same line of lateral thinking as in Mac method is followed, until thinking to an *generate alternative* derivative leads to the new idea of *fractionation* to predict variables at only *half* the time step for *half* grid points $(i, j + 1/2)$ and $(i, j-1/2)$ using Eqs. (25.12–25.13), and similarly for full grid point i, j using Eq. (25.14). Here time derivative at time t is substituted as before from Eq. (25.1) to inject the physics, but with forward differencing proposed for the time derivative in Eq. (25.13), likewise backward differencing for Eq. (25.14) and central differencing for Eq. (25.14).

$$\rho_{i,j+\frac{1}{2}}^{t+\frac{\Delta t}{2}} = \frac{1}{2}\left(\rho_{i,j+1}^{t} + \rho_{i,j}^{t}\right) + \frac{\Delta t}{2}\left(\frac{\partial \rho}{\partial t}\right)_{i,j}^{t} \tag{25.12}$$

$$\rho_{i,j-\frac{1}{2}}^{t+\frac{\Delta t}{2}} = \frac{1}{2}\left(\rho_{i,j}^{t} + \rho_{i,j-1}^{t}\right) + \frac{\Delta t}{2}\left(\frac{\partial \rho}{\partial t}\right)_{i,j}^{t} \tag{25.13}$$

$$\rho_{i,j}^{t+\frac{\Delta t}{2}} = \rho_{i,j}^{t} + \frac{\Delta t}{2}\left(\frac{\partial \rho}{\partial t}\right)_{i,j}^{t} \tag{25.14}$$

Substituting these *better* variables again in Eq. (25.1) (similar to substituting variables with bar on top as in Mac method) leads to the *alternative derivative* sought (represented with a double bar on top), which is used in the final prediction (Eq. 25.16).

$$\left(\overline{\overline{\frac{\partial \rho}{\partial t}}}\right)_{i,j} = -\left(\rho_{i,j}^{t+\frac{\Delta t}{2}}\left[\frac{u_{i,j+\frac{1}{2}}^{t+\frac{\Delta t}{2}} - u_{i,j-\frac{1}{2}}^{t+\frac{\Delta t}{2}}}{\Delta x}\right] + u_{i,j}^{t+\Delta t}\left[\frac{\rho_{i,j+\frac{1}{2}}^{t+\frac{\Delta t}{2}} - \rho_{i,j-\frac{1}{2}}^{t+\frac{\Delta t}{2}}}{\Delta x}\right] + \rho_{i,j}^{t+\frac{\Delta t}{2}}\left[\frac{v_{i,j+\frac{1}{2}}^{t+\frac{\Delta t}{2}} - v_{i,j-\frac{1}{2}}^{t+\frac{\Delta t}{2}}}{\Delta y}\right]\right) \tag{25.15}$$

$$\rho_{i,j}^{t+\Delta t} = \rho_{i,j}^{t} + \Delta t\left(\overline{\overline{\frac{\partial \rho}{\partial t}}}\right)_{i,j} \tag{25.16}$$

Hence lateral thinking gives insight into the genesis of the two methods described.

25.4 Student Feedback

A pilot class covering 'CFD methods with creative thinking' as detailed in this work, was conducted during CFD course 'Computational Fluid Dynamics', taught at Birla Institute of Technology and Science, Pilani, Pilani campus. An anonymous paper based survey gauged the response of the 16 students (Fig. 25.1) registered for the course (January–May 2019). The class comprised eight undergraduate students and seven master's level students which included two female students. The questionnaire appropriated the 5 point Likert scale (Joshi et al. 2015) with six statements (see Table 25.1).

The first three questions were related to the potential of applying lateral thinking specifically to learning CFD, and the next three questions were related to engineering courses in general. It is encouraging to note the positive responses for both sets of questions with slightly more number strongly agreeing with the second set of questions. 85.42% of the responses were either 'agree' or 'strongly agree', 14.58% were 'neutral' and 0% responded with either 'disagree' or 'strongly agree'. The overall-average of all responses was approx. 4, which corresponds to 'agree'.

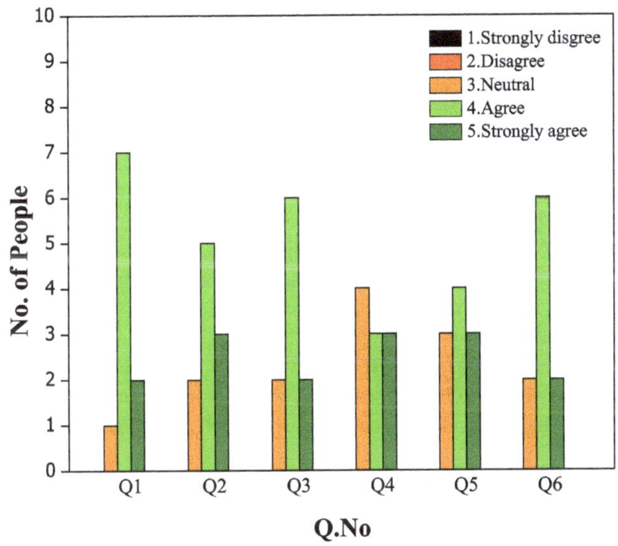

Fig. 25.1 Survey responses of engineering students to the pilot class conducted using lateral thinking to teach CFD methods

Table 25.1 Survey of student feedback

Statements in questionnaire
Q1. I found the techniques of lateral thinking (LT) interesting
Q2. Applying LT techniques to CFD methods provided more insight into the method
Q3. Applying the LT techniques to CFD methods has potential to help me remember them
Q4. Applying the LT techniques has potential to simplify learning for me
Q5. Applying LT may become stress reliving in engineering course, especially when there is so much emphasis on equation based problem solving
Q6. LT techniques have the potential to enhance teaching and learning in other complex and analytical engineering courses as well
Student comments: It may be useful also in optimization and machine leaning type courses; LT techniques helps to remember things which are otherwise difficult; It is a relatively new idea and have the potential to eventually simplify complex concepts for me

25.5 Conclusions

This work explores using lateral thinking in a popular engineering elective such as CFD. Application of Lateral thinking techniques evolved two related CFD methods, from the framework of a basic method. This novel treatment was implemented in a pilot class, as an aid in teaching, learning, simplifying and remembering these complex methods. Survey of student feedback was positive, with all agreeing with the benefits of lateral thinking, not only for CFD, but also potentially for other engineering courses as well.

References

Anderson JD (2012) Computational fluid dynamics: the basics with applications, Indian edn. McGraw Hill, New York

Atwood SA, Pretz JE (2016) Creativity as a factor in persistence and academic achievement of engineering undergraduates. J Eng Educ 105(4):540–559. https://doi.org/10.1002/jee.20130

Conwell JC, Catalano GD, Beard JE (1993) A case study in creative problem solving in engineering design. J Eng Educ 82(4):227–231. https://doi.org/10.1002/j.2168-9830.1993.tb01079.x

Daly SR, Mosyjowski EA, Seifert CM (2014) Teaching creativity in engineering courses. J Eng Educ 103(3):417–449. https://doi.org/10.1002/jee.20048

De Bono E (1970) Lateral thinking: a textbook of creativity. Ward Lock Educational, London

Gordon WJJ (1961) Synectics. Harper & Row, New York

Heilmann G, Korte WB (2010) The role of creativity and innovation in school curricula in the EU27: a content analysis of curricula documents. Joint Research Centre, Luxembourg: European Commission

Joshi A, Kale S, Chandel S, Pal DK (2015) Likert scale: explored and explained. Br J Appl Sci Tech 7(4):396–403. https://doi.org/10.9734/bjast/2015/14975

Klukken PG, Parsons JR, Columbus PJ (1997) The creative experience in engineering practice: implications for engineering education. J Eng Educ 86(2):133–138. https://doi.org/10.1002/j.2168-9830.1997.tb00276.x

Senra M, Fogler HS (2014) Teaching creative thinking and transitioning students to the workplace in an academic setting. Chem En Educ 48(1):9–16

Tekic Z, Tekic A, Todorovic V (2015) Modelling a laboratory for ideas as a new tool for fostering engineering creativity. Procedia Eng 100:400–407. https://doi.org/10.1016/j.proeng.2015.01.384

26

Increasing Employability of Engineers: Courses on Improving Human Skills

Sushila Shekhawat

Abstract Engineering education demands numerous challenges in the present context due to the difficulties posed by the placement scenario. Demand for competent engineering professionals has made the selection process highly challenging giving rise to a need for comprehensive education pedagogy not just confined to inculcating hard core technical attributes but also human skills. Hence there is an urgent need to continuously upgrade the curriculum design of engineering courses so as to better equip the technical graduates with employability skills. BITS, Pilani, one of the premiere technical institutes in India has a well-designed pool of structured courses such as Business Communication, Technical communication, etc. which do incorporate modules such as Professional Presentations, Group Discussions, Interviews, etc. so as to hone the soft skills of the students. The paper attempts to deal with the experiential innovative methodology adopted in the course Business communication so as to enhance the employability skills of the Engineering students.

Keywords Engineering education · Employability skills · Soft skills

26.1 Introduction

The present job scenario demands the workforce to be competitive and well-groomed in all the areas related to hard core technical skills as well as soft skills. "Skills refer to the level of performance of an individual on a particular task or the capability to perform a job well which can be divided into technical elements and behavioral elements" (Noe et al. 2015). Soft skills are the skills, which complement hard skills such as interpersonal communication competence, ability to work in teams, motivation, leadership, etc. As per the recent data LinkedIn's '2019 Global Talent Trends Report' "as automation and AI continue to reshape entire industries, companies and

S. Shekhawat (✉)
Department of Humanities and Social Sciences, Birla Institute of Technology and Science, Pilani - Campus, Pilani 333031, RJ, India
e-mail: sula@pilani.bits-pilani.ac.in

jobs, strong soft skills—the one thing that machines can't replace—are becoming absolutely vital" (Saira 2019).

Hence, comprehensive education curriculum should be the focus of the Educational Institutes in the present scenario. There are organizations whose sole criteria for hiring the professionals is based more on skill sets relevant to a particular profession evaluating an individual on parameters such as attitude, leadership, motivation, team work, persuasive strategies to deal with people, communication skills, emotional intelligence, interpersonal skills, conflict management, etc. However, it is hard to quantify these traits in an individual but certainly adequate training should be provided to inculcate such attributes among the aspiring professionals. Despite the awareness of the expectations of the firms not many educational institutes offer structured courses in the desired area of people skills. There exists a gap between the expectations of the corporate sector and the modules designed by the academic sector. The growing trend of soft skills training sometimes fail in instilling an urgency among engineering students as they still consider completing the basic educational requirement as the benchmark. Büth et al. (2017) quote the AICTE, Placement, Graduates, Enrolments and Intake Statistics of Engineering Education in India.

"The All India Council for Technical Education (AICTE) statistics show that the enrolment of youth for technical courses in India is lower than the available capacity and less than 40% of the graduating students get jobs in industry. This clearly shows a mismatch between the industry requirements and academic skill development" (Büth 2017).

However, a few educational institutions include such courses as essential components in their academic structure and BITS Pilani being one of the leading educational institutions in India place a lot of importance to such courses. The course business communication taken up for analysis in this paper is a well-established course initiated in the year 2002 and since then it has been a highly popular course among the student community. The main objective of the paper is to share the teaching modules included in the course structure and the positive impact on the learning outcome of the students.

In BITS Pilani, several courses dealing with soft skills have been the essential part of the education curricula such as business communication, technical communication, etc. All these courses have all the desired soft skills components in their structure where students acquire significant training in all the people skills relevant to their job profile. Though all these courses have their specific objectives, this paper would confine its discussion to only Business communication.

26.2 Literature Review

Hard core skills are the skill sets relevant to a particular discipline but soft skills are a set of traits required for imparting a wholesome education to a professional wherein he is not only a computer engineer or a mechanical engineer but a technical graduate with a perfect hold on his dealings with the professional world and hence can be a good leader, a motivator and a person having all the necessary attributes which enable him to deliver his best in the job scenario. As defined by Hargis, technical skills are also termed as hard skills, and are "job specific tasks directly necessary for successful completion of the job," for example, electricity, robotics, and computer technology (Hargis 2011).

Various studies have been conducted which highlight the role of soft skills also known as Employability skills for a technical professional and Chaita explains employability skills as "thinking skills such as logical and analytical reasoning; problem solving; capacity to identify access and manage knowledge and information; personal attributes such as imagination; creativity and intellectual rigor; values such as ethical practice, persistence, integrity and tolerance, problem solving, team working, communication, leadership" (Chaita 2016).

On the other hand, another scholar Dorsey compares employability skills with medicine which one has to gulp in order to survive in the job market. He highlights the vital nature of employability skills by terming them to be unavoidable and thinks that they "are often times the area that will determine failure or success for many in the workplace in personal life as well as individual contractor. Some of the most important people in the world are failing, because though they have mastered the hard skills, they can't cope with the world that is fuelled by soft skills." (Dorsey 2004).

In addition to the above stated perspectives on employability skills it has also been defined by Keller, Parker, and Chan as an assorted array of knowledge, skills, and attributes that are relevant for the workplace (Keller et al. 2011). Employability skills include two categories of skills: technical and soft skills (Omar et al. 2012).

However, to inculcate these skills among the students an interactive methodology has been adopted which generates their interest and enable them to gain proficiency in soft skills.

26.3 Methodology

The main objective of the course Business communication is to provide training to the students in both the oral and written communication skills, necessary for various managerial activities such as conducting and participating in interviews, discussing in groups, presenting individually and in teams, speaking in public, giving instructions, conducting meetings, etc. The components included in the course along with the proper interactive delivery in classrooms enable the students to understand and demonstrate the use of proper and advanced writing techniques that today's technology demands including anticipation and audience reaction. As it is a course offered every second semester, so the methodology, case studies and the assessments vary in order to bring in a fresh perspective every time.

The course has several components such as initial lectures based on providing exposure to different kinds of communication activities in an organization so as to better equip the learners with the nuances associated with the significance of different kinds and modes of communication. The objective of the case studies adopted for the communication module is to create an awareness among the students about the professional working environment and it establishes a base for learning the soft skills.

Case study 1: After giving them theoretical inputs through certain examples, the students are asked to visit different work segments of BITS Pilani Library and come up with all the essential aspects related to interpersonal communication associated with the working of Library. The same case study is evaluated as an assignment where the focus is on the following skills:

- Motivation to work in teams
- Attending team sessions and
- Delivering Presentation
- Handling questions and answers.

In addition to teaching the essentials of communication for managers' students are encouraged to come up with their own presentation topics and all the topics are further discussed in the class to avoid repetitions. In this exercise, they develop the skills to discriminate between the important and not so important topics and after a lot of deliberation among the team members a final decision about the topic is taken. The learners are also encouraged to participate openly during classroom lectures so as to provide them a friendly ambience and allow free exchange of information on any topic. The different soft skills teaching modules are dealt with separately as discussed in the upcoming sections.

26.3.1 Professional Presentations

During their professional presentations certain yardsticks are provided to them regarding the following parameters:

1. organization of content
2. different delivery modes
3. significance of engaging the audience
4. specifics of non-verbal communication
5. time management.

At the end of every presentation, instructor shares the feedback and marks are kept for students' involvement and handling questions from the audience. The presentations are individual presentations or team presentations depending upon the strength of the students.

26.3.2 Group Discussion

It is another important component included in this course wherein topics ranging from general, abstract to the more specific ones related to any current event are included. In order to make students feel more involved students are advised to provide their own topics of interest. In-depth knowledge with an accurate display of a range of original ideas during group discussion is a desirable trait evaluated in a participant. Analytical ability, logical thinking, ability to convince people of one's ideas and making them agree to your desired notions is an indispensable characteristic tried and tested in group discussions. Hence, the students are made to go through a mock group discussion first so as to know the kind of level they have and their degree of involvement in the team as a member and also knowing the skill to showcase contradiction, agreement

Persuasion in work place situations is another significant aspect student should develop during the entire group discussion process. Group discussion is introduced to them as an activity not just significant for their placement process but also as an integral aspect of the organizational growth wherein they have to deal with different kinds of issues and hence are adequately trained to have qualities of originality, innovation, conviction, confidence and maturity. Responsible way of solving problems with workable solutions and decision making skills are some of the traits students develop through these group discussion exercises.

At the end of the group discussion, they are asked to rate themselves on distinct parameters and the same kind of feedback is taken from the audience who are asked about the loopholes of the group discussion on both the topic being discussed and the kind of delivery of the participants. Here also lot of interaction takes place among the entire class which makes the students feel not just passive recipients but active contributors allowed to offer their own insight into the topic being discussed.

26.3.3 Interviews

Another vital module, which is very well handled in the course through less theoretical content and more practice based activities. In the initial interview sessions, the students are divided in groups and made to study on their own and through interaction with their seniors who are undergoing placement interviews. In the class, they are divided in groups of five to six students depending upon the strength of the class and asked to write on a sheet of paper the possible questions that can be asked in an interview and think of some possible answers. Entire lecture sessions are made interactive and theoretical concepts are very well integrated with the possible areas and the probable answers. In the next step, they are made to undergo a rigorous personal interview session in the instructor's chamber and two more subject experts are invited to form a panel.

In order to bring seriousness among the students they are asked to dress up formally, prepare complete Curriculum vitae and come for the interview. The criteria for evaluation is based on the parameters is based on the parameters such as:

- Personality
- Knowledge
- Communication skills
- Leadership skills
- Problem solving skills
- Decision making skills.

Each student is allotted fifteen to twenty minutes for the interview round and questions are asked by the panelists including the instructor to test the above stated traits. At the end of all the interviews, the feedback of the panelists is shared with the students and the marks are announced along with the areas of strength and improvement. The students are asked about their experiences about the interview component and their possible suggestions are also noted by the instructor.

26.4 Results and Conclusion

Feedback from the students after every session is considered mandatory which lets the instructor test the knowledge acquired during the entire course. Out of 80 students registered in the course at least 75% rate the course to be successful in inculcating the soft skills among the entire population. On being asked about the components they considered the most important and interesting ones some students rate the components differently but Group discussions and interviews are being rated as the most important skills due to the urgent requirement in their placement process.

However, there are certain differences in terms of the fluency in expression and the language as there are certain students who are not at par with the mainstream engineering students due to their educational background and hence their performance is not as good as the engineering students.

Practice based components and the interactive methodology enables the students to learn better and provide them an opportunity to not only learn through doing but also gets an exposure in understanding the indispensability of the skills which is considered to be quite insignificant during their college years and can be learned later. As the students in BITS Pilani are highly proficient in their language they are also made to learn the difference between language proficiency and the honing of skills required for their future job. Realization of this urgent need of human skills prompts them to work hard and participate not just during classroom sessions but even after lecture hours to discuss the possible strategies to further improve these skills. However, with the changing classroom, teaching scenario, which should be more of practice oriented rather than merely confined to delivering lectures there is a need to continuously update the teaching pedagogy to be more and more interactive and task based.

References

Büth et al (2017) Bridging the qualification gap between academia and industry in India. Proc Manuf 9:275–282

Chaita MV (2016) Developing graduate employability skills: your pathway to employment. Universal, Boca Raton, USA

Dorsey I (2004) Soft skills for hard times: a handbook for high achiever. Xlibris Corporation, United States of America

Hargis KB (2011) Career and technical education program alignment with local workforce needs (Ed.D. dissertation). Eastern Kentucky University (UMI No. 3488204)

Keller S, Parker CM, Chan C (2011) Employability skills: student perceptions of an IS final year capstone subject. ITALICS 10(2):4–15

Noe RA, Hollenbeck JR, Gerhart B (2015) Fundamental of human resource management. McGraw-Hill, New York

Omar MK, Bakar AR, Rashid AM (2012) Employability skill acquisition among Malaysian community college students. J. Soc. Sci. 8(3):472–478. https://doi.org/10.3844/jssp.2012.472.478

Saira A (2019) The friendly hire: why soft skills are essential in the workplace. https://www.thehindu.com/education/careers/the-friendly-hire-why-soft-skills-are-essential-in-the-workplace/article26408089.ece. Accessed 18 Dec 2019

27

Digitalization, Mergers and Acquisitions

Leonhard Riedel and Reza Asghari

Abstract Digitalization requires business development agents to align their business models to meet customer demands and competition requirements. To achieve this transformation however is a challenge for incumbent organizations, since over time they have often developed rigidity and they might suffer under lack of necessary competencies. Acquiring digital start-ups is an often used and promising strategy to close these feasibility gaps. However, digitalization and mergers and acquisitions have not yet been brought together in business sciences. Therefore, this paper aims on identifying process steps of both, digital business transformation and mergers and acquisition models and to relate these process steps for designing a new integrated process. Finally, the resulting process is described in terms of main tasks and their sequence. The paper concludes that future research should focus on integrating operative methods to provide further support for incumbents when facing digitalization.

Keywords Digitalization · Mergers and acquisitions · Business transformation

27.1 Introduction

Technological achievements are a major driver of organizational change. There are various ways in which technological progress can affect innovating firms. It can lead to new opportunities for product innovation, value creation or business models. However, opportunities and risks often arise at the same time. New competitors can enter the markets and changing customer requirements must be understood and met (Zainun and Smith 2011).

Innovating incumbencies are often confronted by two main challenges relating technological progress. On the one hand, they are required to establish a suitable knowledge fundament within the organization as a pre-requirement for innovation.

L. Riedel (✉) · R. Asghari
Entrepreneurship Hub, Technische Universität Braunschweig, 38106 Braunschweig, Germany
e-mail: Leonhard.Riedel@tu-braunschweig.de

On the other hand, they need to harmonize the firm's managerial system of coordination and control according to these technology-related external changes (Pavitt 2005). Especially today, in times of ever accelerating technological progress, this means a severe challenge for established enterprises (Peter 2017). Digitalization, a term used to describe the managerial issues that emerge by new digital technologies, requires managers to find the ideal organizational configuration. This is achieved by multifaceted change processes, often referred to as digital business transformation (DBT) (Obermaier 2016; Bowersox et al. 2005; Picot et al. 2017). DBT can provide benefits, such as increasing customer satisfaction and loyalty, improving process efficiency or creating new revenue streams. Corporate organizations are entailed to carry out DBT to maintain their competitiveness (Teece and Linden 2017). Yet, DBT requires certain digital competencies and business agility that both often lack in incumbent organizations (Augsten et al. 2017; BMWi 2016). While only a few incumbents have these characteristics, they are particularly pronounced in digital start-ups (Kreutzer and Land 2013). Collaboration and alliances among established firms and digital start-ups thus can be a promising strategy. While incumbencies profit by overcoming their feasibility gaps by acquiring lacking digital technologies and competencies, start-ups can benefit mainly by access to lacking resources (Leube and Grellier 2017; Islam et al. 2017).

There are several forms of how incumbencies can cooperate with start-ups. Mergers & Acquisitions (M&A) is the most intense and financially riskiest form of cooperation. Yet, it allows to attain the entire technology stock and competence properties of a start-up and consequently enables for the most radical DBT in a short time (Ernst and Young 2019; Feix 2018). In practice, organizations are already using M&A activities for DBT. In literature however, DBT and M&A have yet only been subject of distinct research fields. This paper attempts to close the gap by suggesting a procedure that includes phases of DBT and M&A. Enterprises benefit by our findings as our approach helps navigating along the DBT process and highlights M&A as a feasible strategy to foster DBT capabilities. The paper aims on a consistent and harmonized process. The resulting process addresses both key challenges of technology induced organizational change mentioned above. First, the development of a required knowledge foundation to carry out DBT by acquiring required competencies. Secondly, the M&A integrated DBT process supports organizations to coordinate and control organizational change by providing a systematic guideline.

To achieve these goals, Sect. 27.2 presents process steps of DBT and M&A as a foundation for the integrated process. In that section, several process steps are derived from two literature reviews. Section 27.3 allocates M&A and DBT process steps and provides a step by step description of the integrated process. The paper closes by a brief discussion and conclusion.

Table 27.1 Review on process steps of DBT

References	Process steps					
Bloching et al. (2015)	Analyze digitalization effects on industry	Reflect on corporate situation	Implementation roadmap			
Carolis et al. (2017)	Maturity assessment	Strengths and weaknesses	Opportunities Identification	Digital roadmap		
Deloitte and Touche GmbH Wirtschaftsprüfungsgesellschaft (2019)	Sensitization	Implementation	Transformation	Continuous innovation		
Esser (2019)	Analysis	Strategy	Design	Impact	Transformation	
Schallmo (2016)	Digital reality	Digital ambition	Digital opportunities	Digital fit	Digital implementation	
Botzkowski (2018)	Analyze imminent disruption	Business model and value chain status	Transformation Goals	Check available resources	Business transformation	Evaluation and learning
Issa et al. (2018)	Task force setup	Digitalization assessment	Focus definition	Use case idea generation	Use case impact estimation	Use case selection

27.2 Elements of DBT and M&A

27.2.1 Process Steps of DBT

This section provides an overview of different DBT models that have been described in literature. In our study we solely include DBT models that distinguish different process steps. We identify seven models of DBT with a range of three to six process steps. The resulting compendium of DBT process steps is shown in Table 27.1.

From the literature review it can be concluded that DBT models differ in terms of their process steps and scope. Some authors describe personnel-related initialization phases. They emphasize the need for sensitization and the formation of a task force. However, most authors begin the DBT process with an analysis of technology-related changes. These changes are then mapped against the overall organizational situation and capabilities to form a business transformation strategy. The overall scopes of DBT processes differ in that three contributions close with the DBT implementation plan, two with the implementation itself and two go beyond the actual implementation by learning or a continuation of innovation and change.

27.2.2 Process Steps of M&A

In this section, M&A process steps from literature are accumulated and compared with each other. Similar to our approach regarding the analysis of DBT phases, we

Table 27.2 Review on process steps of M&A

References	Process steps			
Feix (2017)	M&A strategy	Transaction management		Integration
Frankenberger et al. (2006)	Pre-merger	Merger		Post-merger
Weber (2013)	Planning strategic management	Negotiation, due diligence and agreement		Integration
Jansen (1998)	Strategic analysis and conception	Transaction	Integration	Monitoring
Wöhler and Cumpelik (2006)	Strategy	Screening	Transaction	Integration

only include procedural models of M&A that distinguish distinct process steps. A resulting compendium of M&A process steps is provided in Table 27.2.

The results indicate that M&A processes start with an initialization phase in which the incumbent reviews the strategic business environment and identifies a strategic gap. On the basis of these objectives, an M&A strategy is then derived, and a screening is carried out to identify suitable business units. After successful negotiation, the external business is integrated into the company. Two procedural models suggest phases beyond integration by reviewing and monitoring goal achievements.

27.3 Introducing the M&A Integrated DBT Process

27.3.1 Convergence and Complementarity of M&A and DBT

To propose an M&A integrated process for DBT it first requires a conclusive and deeper understanding for both individual processes. This is realized by separately connecting the individual process steps in Sects. 27.2.1 and 27.2.2. Then, it is analyzed how the M&A and DBT processes are related to one another, i.e. which process steps share similar contents. Finally, process steps of M&A that cannot be integrated into existing process steps of DBT are allocated to suitable positions of the DBT process. The results are shown in Fig. 27.1. The upper arrow describes the M&A process and the lower arrow describes the DBT process. Process steps with equivalent content that can be merged are referred to as convergent. M&A process steps that cannot be integrated into existing DBT process steps are termed complementary and become additional steps for the M&A integrated DBT process. The merging and assignment of complementary and convergent M&A process steps to suitable positions in the DBT process is visualized by the arrows in the allocation section. The results show that M&A can be integrated into the DBT process. The M&A strategy must be derived from the individual problem statement of the established enterprise. This in turn is based on a specific analysis of the business environment. Thereby,

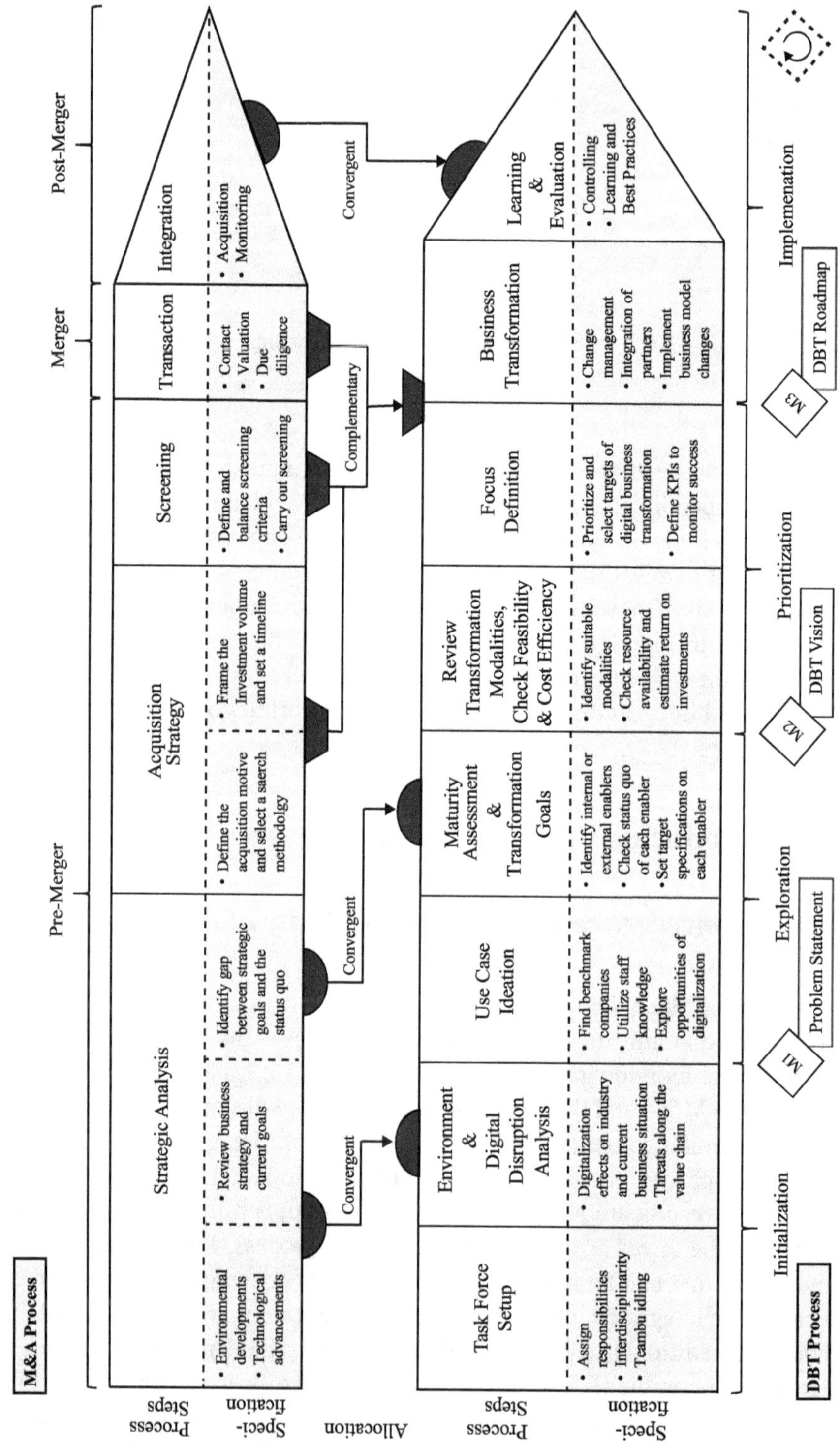

Fig. 27.1 Integrating the M&A process into the DBT process

analysis-related process steps of DBT focus more on technology-driven opportunities and risks than the general M&A process. In addition, DBT includes a more distinctive creative exploration of ideas and strategies to cope with these external changes. The resulting M&A integrated DBT process is described more thoroughly in Sect. 27.3.2.

27.3.2 Description of the Integrated Process

This section outlines main activities of different process steps of M&A integrated DBT. The overall process has been divided into nine steps in four phases with distinct aims and purposes.

Initialization Phase. The Initialization Phase aims at mobilizing forces of organizational change and building a shared understanding of emerging threats.

Task Force Setup. DBT concerns the whole organization. Therefore, in the first step, an interdisciplinary task force is set up. It requires interdisciplinarity to carry out multifaceted discussions from different perspectives on the organization and to provide detailed and relevant information regarding all organizational departments (Issa et al. 2018). A balanced composition of the Task Force is therefore crucial for a successful DBT.

Business Environment Analysis and Digital Disruption Analysis. In second step the overall business environment must be reviewed. The task force analyzes the status quo of the organization and its strategy (Jansen 1998). Emergent changes in the industrial sector due to digitalization are discussed. Thereby, the task force not only focuses on the organization itself, but also on digitalization effects on the whole value chain (Bloching et al. 2015). They review on changing customer demands and how digitalization may aggravate the competitive situation (Esser 2019; Schallmo 2016; Botzkowski 2018; Jansen 1998).

Exploration Phase. In this phase, the task force develops a digitalization vision for the company. They also gain a deeper understanding of how this vision can be achieved by identifying related enablers and examining their status.

Use Case Ideation. The first step of the exploration phase is to identify benefits of digitalization and to generate broad ideas of what can be implemented (Schallmo 2016). Hence, the task force identifies best practices from other organizations in different areas of the business model or utilizes internal staff knowledge for ideation (Deloitte and Touche GmbH Wirtschaftsprüfungsgesellschaft 2019; Issa et al. 2018).

Maturity Assessment and Transformation Goals. In this consecutive step, organizational enablers that contribute synergistically to each use case are identified (Schallmo 2016). The maturity levels of these enablers are then determined (Carolis et al. 2017; Esser 2019). In the case of gaps, between the status quo and requirements the closure of these gaps is set as a goal of DBT when following respective use cases (Deloitte and Touche GmbH Wirtschaftsprüfungsgesellschaft 2019; Botzkowski 2018).

Prioritization Phase. The two process steps in this phase support companies in defining a digitization roadmap. This strategic plan contains specific objectives of what is to be changed first to transform the organization and how this transformation is to be carried out.

Review Transformation Modalities, Check Feasibility and Cost-efficiency. In this step, the task force first evaluates various modalities of carrying out DBT. The main questions are whether the company has the capabilities and resources to fulfill parts of the digitization vision and how external resources could support this process (Esser 2019; Schallmo 2016; Botzkowski 2018). These DBT modalities will then be subject to feasibility and cost-efficiency evaluations (Issa et al. 2018).

Focus Definition. After the task force has reviewed alternative elements of the digitalization vision and their modalities, these options are then prioritized to identify the ideal DBT starting point (Bloching et al. 2015; Carolis et al. 2017; Schallmo 2016; Botzkowski 2018).

Implementation Phase. This phase contains the implementation of DBT via M&A.

Acquisition Strategy and Screening. In case M&A is an attractive option for closing the factual gaps to carry out DBT, an acquisition strategy is drafted based on the digitalization roadmap. It frames the overall M&A purpose and activities and serves as a starting point for defining screening criteria (Feix 2017; Weber 2013; Jansen 1998; Wöhler and Cumpelik 2006). The task force defines search methods and completes the screening. This process step results in identifying suitable digital start-ups to be addressed.

Transaction, Integration and Business Transformation. Once both parties come to an agreement, the start-up is integrated into the incumbent organization (Feix 2017; Frankenberger et al. 2006; Weber 2013; Jansen 1998; Wöhler and Cumpelik 2006). Therefore, organizational change management and integration management converge to carry out DBT. Goals that have been set up within the Prioritization Phase are then operationalized. Related changes of the business models are tested on the market (Deloitte and Touche GmbH Wirtschaftsprüfungsgesellschaft 2019; Esser 2019; Schallmo 2016; Botzkowski 2018).

Learning and Evaluation. Finally, it is examined whether the formal objectives of the digital transformation have been achieved and which empirical findings can be gained from the DBT experience for future business development or M&A activities (Botzkowski 2018; Jansen 1998).

With an ever-changing business environment, the overall DBT process is iterated, leading to a continuous and unceasing business development (Deloitte and Touche GmbH Wirtschaftsprüfungsgesellschaft 2019).

27.4 Discussion and Conclusion

By introducing a new process of DBT that integrates M&A, the paper provides a consistent process for closing feasibility gaps of incumbents when confronted with digitalization. The main limitation for this paper is that the interlocked relationships of M&A and DBT processes have not been regarded in an applicable perspective, such as how screening criteria can be derived and balanced based on the business environment analysis. Also, digitalization may affect the whole M&A process, such as start-up data bases or technological tools to support the post-merger integration. Hence, the digitalization effects on the M&A process itself can be helpful to provide further support for organizations. Future research can also focus on operationalizing DBT by integrating appropriate tools and methods into the process provided in this paper.

References

Augsten T, Brodbeck H, Birkenmeier B (2017) Strategie und Innovation: Die entscheidenden Stellschrauben im Unternehmen wirksam nutzen. Springer Gabler, Wiesbaden (2017). https://doi.org/10.1007/978-3-658-15684-8

Bloching B, Leutiger P, Oltmanns T, Rossbach C, Schlick T, Remane G, Quick P, Schafranyuk O (2015) Die digitale Transformation der Industrie: Was sie bedeutet. Wer gewinnt. Was jetzt zu tun ist. Roland Berger Strategy Consultants GmbH, Bundesverband der Deutschen Industrie e.V. (eds) München

BMWi (2016) Aktionsprogramm Zukunft Mittelstand: Neuauflage 2016. Bundesministerium für Wirtschaft und Energie (eds) Berlin

Botzkowski T (2018) Digitale Transformation von Geschäftsmodellen im Mittelstand: Theorie, Empirie und Handlungsempfehlungen. Springer Gabler, Wiesbaden. https://doi.org/10.1007/978-3-658-20333-7

Bowersox DJ, Closs DJ, Drayer RW (2005) The digital transformation: technology and beyond. SCMR 9(1):22–29

Carolis A de, Macchi M, Negri E, Terzi S (2017) Guiding manufacturing companies towards digitalization: a methodology for supporting manufacturing companies in defining their digitalization roadmap. In: ICE/IEEE ITMC, Piscataway, New Jersey, pp 487–495. https://doi.org/10.1109/ice.2017.8279925

Deloitte & Touche GmbH Wirtschaftsprüfungsgesellschaft (2019) Überlebensstrategie "Digital Leadership". https://www2.deloitte.com/content/dam/Deloitte/at/Documents/strategy/ueberlebensstrategie-digital-leadership_final.pdf. Accessed 06 July 2019

Ernst & Young (2019) Digital transformation for 2020 and beyond: a global telecommunications study. https://www.ey.com/Publication/vwLUAssets/ey-digital-transformation-for-2020-and-beyond/$FILE/ey-digital-transformation-for-2020-and-beyond.pdf. Accessed 07 June 2019

Esser M (2019) Chancen und Herausforderungen durch Digitale Transformation. http://www.strategy-transformation.com/digitale-transformation-verstehen/. Accessed 07 June 2019

Feix T (2017) Das M&A-Prozessmodell für den Mittelstand. In: Feix T, Büchler J-P, Straub T (eds) Mergers & Acquisitions: Erfolgsfaktoren für mittelständische Unternehmen. Haufe, Freiburg, pp 15–33

Feix T (2018) Digitalisierung, business model innovation und M&A. M&A Rev 29(10):336–343

Frankenberger S, Kappler F, Müller-Stewens G (2006) Organisationsentwicklung in M&A Prozessen: Gruppe. Interaktion. Organisation. GIO 37(1):67–78. https://doi.org/10.1007/s11612-006-0009-9

Islam N, Buxmann P, Ding D (2017) Fostering digital innovation through inter-organizational collaboration between incumbent firms and start-ups. In: ECIS 2017 proceedings. Research papers, Guimares, pp 1029–1043 (2017)

Issa A, Hatiboglu B, Bildstein A, Bauernhansl T (2018) Industrie 4.0 roadmap: framework for digital transformation based on the concepts of capability maturity and alignment. Proc CIRP 72:973–978. https://doi.org/10.1016/j.procir.2018.03.151

Jansen SA (1998) Mergers & acquisitions. Springer Gabler, Wiesbaden. https://doi.org/10.1007/978-3-8349-4772-7_1

Kreutzer RT, Land KH (2013) Digitaler Darwinismus. Springer Gabler, Wiesbaden. https://doi.org/10.1007/978-3-658-08547-6

Leube HP, Grellier A (2017) High-Tech Companies: Auswirkungen der digitalen Disruption auf M&A Aktivitäten. M&A Rev 28(4):123–127

Obermaier R (2016) Industrie 4.0 als unternehmerische Gestaltungsaufgabe: Strategische und operative Handlungsfelder für Industriebetriebe. In: Obermaier R (ed) Industrie 4.0 als unternehmerische Gestaltungsaufgabe. Springer Gabler, Wiesbaden, pp 3–34. https://doi.org/10.1007/978-3-658-08165-2_1

Pavitt K (2005) Technologies, products and organization. What Adam Smith tells us and Joseph Schumpeter doesn't. In: Dosi G, Teece DJ, Chytry J (eds) Understanding industrial and corporate change. Oxford University Press, Oxford, pp 267–285 (2005). https://doi.org/10.1093/icc/7.3.433

Peter MK (2017) KMU-transformation. Als KMU die Digitale Transformation erfolgreich umsetzen: Forschungsresultate und Praxisleitfaden. FHNW Hochschule für Wirtschaft, Olten

Picot A, Hopf S, Sedlmeir J (2017) Digitalisierung als Herausforderung für die Industrie: Das Beispiel der Automotive Branche. In: Burr W, Stephan M (eds) Technologie, Strategie und Organisation. Springer Gabler, Wiesbaden, pp 87–112. https://doi.org/10.1007/978-3-658-16042-5_5

Schallmo DRA (2016) Jetzt digital transformieren: So gelingt die erfolgreiche Digitale Transformation Ihres Geschäftsmodells. Springer Gabler, Wiesbaden. https://doi.org/10.1007/978-3-658-23409-6

Teece DJ, Linden G (2017) Business models, value capture, and the digital enterprise. JOD 6(1). https://doi.org/10.1186/s41469-017-0018-x

Weber Y (2013) A comprehensive guide to mergers & acquisitions: managing the critical success factors across every stage of the M&A process. FT Press, New Jersey

Wöhler C, Cumpelik C (2006) Orchestrierung des M&A-Transaktionsprozesses in der Praxis. In: Wirtz BW (ed) Handbuch mergers & acquisitions management. Springer, Heidelberg, pp 453–467

Zainun TT, Smith M (2011) The effects of changes in competition, technology and strategy on organizational performance in small and medium manufacturing companies. Asian Rev Account 19(3):208–220. https://doi.org/10.1108/13217341111185137

28

Women Entrepreneurs and Entrepreneurial Intention: Impact of Motivation

Sakshi Chhabra, Rajasekaran Raghunathan, and Navuluru Venkata Muralidhar Rao

Abstract The purpose of this paper is to explore the key factors, which motivate women to take up entrepreneurial activity and to understand its role in increasing Entrepreneurial intention towards women entrepreneurs in Indian micro, small and medium enterprises. This paper attempts to develop a scale for measuring motivational factors, entrepreneurial potential that leads to entrepreneurial intention among women entrepreneurs in India and hence testing its validation. In addition, it will help in understanding the relationship between motivation, entrepreneurial potential and intention through hypothesis testing. The paper adopts an exploratory and descriptive research design capitalizing on authentic and reliable secondary data through exhaustive studies of reputed journals/literatures, government sources. A sample of 397 respondents from all across the India has been collected from women entrepreneurs for this study in order to understand the motivational factors and its role in increasing entrepreneurial intention, using cluster and snowball sampling on self-administered questionnaire. The accumulated data were than analyzed using descriptive analysis for validity and reliability checks. Strong correlations were found between motivational factors, perceived desirability, entrepreneurial potential and intention and hence confirmed that all the measures in the instrument were well constructed. Hypothesis testing using Spearmen's correlation test (two tailed) explains the significance of relationship between the variables.

S. Chhabra (✉)
Department of Management, Birla Institute of Technology and Science, Pilani, Pilani-Campus, Pilani 333031, RJ, India
e-mail: p2014011@pilani.bits-pilani.ac.in

R. Raghunathan
Department of Economics and Finance, Birla Institute of Technology and Science, Pilani, Hyderabad-Campus, Hyderabad 500078, TN, India
e-mail: raghu@hyderabad.bits-pilani.ac.in

N. V. M. Rao
Department of Economics and Finance, Birla Institute of Technology and Science, Pilani, Pilani-Campus, Pilani 333031, RJ, India
e-mail: nvmrao@pilani.bits-pilani.ac.in

Keywords Women entrepreneurship · Motivation · Entrepreneurial intention

28.1 Introduction

The twenty-first century has witnessed a paradigm shift in the role of women from being simply a home maker to a career oriented professional. This in turn led to improvement in their economic strength as well as position in society. Having realized this present day, modern women do not want to confine themselves within the boundaries of their houses. Women entrepreneurship is an act of owning a business, which makes women economically independent. It has been found that an estimated 126 million women were starting or managing new businesses in 67 countries around the world. In addition, an estimated 98 million were running established businesses. Globally men make up 52% of all entrepreneurial activity and remaining 48% of them constituted women entrepreneurs. The gender gap in entrepreneurial activities varies across the world. It ranges from 1.5 to 45.4% women of the adult population (Kelley 2012).

Women entrepreneurship is considered to be instrumental in women's empowerment and improvement in their quality of life. Entrepreneurial activity originates at the individual level and can always be traced back to a single person, the entrepreneur. Entrepreneurship is, hence, induced by an individual's attitudes or motives, skills and psychological endowments. The gender gap in entrepreneurship is defined as the difference between men and women in terms of numbers engaged in entrepreneurial activity, motives to start or run a business, industry choice and business performance and growth. Most entrepreneurship research on individual level analysis has focused on the entrepreneurial activities of male entrepreneurs. Few Researchers in their study has criticized earlier research findings stating that there is no significant difference in male and female entrepreneurs (Carter and Cannon 1988). Several other studies were conducted with a focus on studying the gender differences in entrepreneurship. It has been observed through the studies that women entrepreneurs are different from that of men in terms of characteristics, background, motivation, entrepreneurial skills and the problems faced by them (Hisrich and Candida 1984; Lituchy and Reavley 2015; Bird and Brush 2002).

Global Entrepreneurship and Development Institute (GEDI) conducted a study and ranked India at 70 among 77 countries. In this study, India has obtained a score of 25.3 in the Female Entrepreneurship Index and the study suggested a need for significant changes to reduce the barriers for female entrepreneurs in India. This study further identified that, women entrepreneurs in India were more opportunity driven than necessity driven (Terjesen and Lloyd 2015). BNP Paribas Global Entrepreneurialism Report has ranked India on top for the highest percentage of successful women entrepreneurs. Nearly half (49%) of successful entrepreneurs in India are female entrepreneurs. A majority of these businesses are in the micro sector (BNP Paribas 2015). One of the major study highlighted that micro, small and medium enterprises (MSME's) have been recognized as a fundamental driver for creating, running and growing business and consequently the economic growth of a country and this is no different for Indian economy as well (Ascher 2012). As mentioned in International finance corporation, world Bank and small and medium enterprises report (IFC Issue Brief/Small and Medium Enterprises 2010), globally MSME's constitutes about 90% of the business enterprises. This research intends to study the various motivational factors and clarifies the fact that how these factors influence the entrepreneurial intention of women entrepreneurs. The focus is also made to understand the entrepreneurial intention among women in particular studying the motivational factors that leads to entrepreneurial potential in women entrepreneurs.

28.2 Literature Review

Most entrepreneurship research on individual level analysis has focused on the entrepreneurial activities of male entrepreneurs. Rapid increase in the number of women owned enterprises across different countries has attracted several researchers' interest. There are not much attempts made to conduct a nation-wide comprehensive research study in order to understand the motivation, issues and challenges of women entrepreneurs. Most of the research studies are cross sectional or longitudinal studies focusing on specific regions within India, which includes Tamilnadu, Jammu and Kashmir, Meghalaya, Andhra Pradesh, Karnataka, Uttar Pradesh and Kerala (Marichamy 2013; Mani 2011; Thomas and Lavanya 2012; Mustafa 2013; Shah and Mustafa 2014; Kurbah 2013; Nagalakshmi 2015; Tarakeswara and Tulasi 2013; Shiralashetti 2014; Dwivedi and Mishra 2013; Chandrasekar et al. 2008).

An attempt is made in this study to fill these literature gaps. Hence, the literature highlights the various motivational factors and the role of these factors in influencing entrepreneurial intention among women. The comprehensive review of literature has been divided into three sections: defining Women entrepreneurship, motivational factors and its role in increasing Entrepreneurial Intention.

28.2.1 Defining Women Entrepreneurship

Women entrepreneurship is a subset of entrepreneurship; it also faces the difficulty of single or commonly accepted definition for the same. Researchers have defined women entrepreneurship in several ways; Table 28.1 summarizes some of these definitions.

Definition of micro, small and medium enterprises was given under Micro, small and medium enter-prises development (MSMED) Act, 2006 is considered for this study. This act defines women entrepreneurship as "a business entity which owned and managed by a woman" (Ahl 2006). It was in 1970s that the Government of India began to promote and focus on self-employment among women. Because of these

Table 28.1 Selected definitions of women entrepreneurship

Author(s)	Definition of women entrepreneurship
Vinze (1987)	A woman entrepreneur is a person who is an enterprising individual with an eye for opportunities and an uncanny vision, commercial acumen, with tremendous perseverance and above all a person who is willing to take risks with the unknown because of the adventurous spirit she possesses
Anwar and Rashid (2012)	A woman entrepreneur can be defined as a confident, innovative and creative woman capable of achieving self-economic independence individually or in collaboration, generates employment opportunities for others through initiating, establishing and running the enterprise by keeping pace with her personal, family and social life
Sharma (2013)	Female entrepreneurs are defined as those who use their knowledge and resources to develop or create new business opportunities, who are actively involved in managing their businesses, and own at least 50% of the business and have been in operation for longer than a year
Government of the People's Republic of Bangladesh (2010)	Woman or a group of women who initiate, organize and run a business enterprise. Women entrepreneur is any women who organizes and manages any enterprise, usually with considerable initiative and risk
Rummana (2014)	A woman will be termed as a Woman Entrepreneur if she is the 'owner or proprietor of a private or proprietary enterprise' or 'is the director of a private company' registered with the 'joint stock' or 'shareholding enterprise' or owning at least 51% share among the shareholders'
Masood (2011)	An enterprise owned and controlled by a women having a minimum financial interest of 51% of the capital and giving at least 51% of the employment generated in the enterprise to women
Singh and Raina (2013)	A woman will be termed as an Entrepreneur if she is the 'owner/proprietor/director of a private/proprietary enterprise/private company' registered with the 'joint stock' or 'shareholding enterprise', owning at least 51% annual turnover and share among the shareholders' and generates employment opportunities for others by administering the enterprise

initiatives, in late ninety's the concept of women entrepreneurship gained prominence. Women owned enterprises account for 25% of them. American Express OPEN report on 'State of women owned business' ranks India at 16th in terms of revenue generation by women owned enterprises. In 2009, a study carried out by Centre for women's business research indicates women entrepreneurs in India were growing twice as fast as the other business (Mallya 2012). After an extensive analysis of these studies, the following key themes emerged background, profile, characteristics, motivation and entrepreneurial Intention.

28.2.2 Women Entrepreneurship and Motivation

Entrepreneurial Motivation is the drive of an entrepreneur to maintain an entrepreneurial spirit in all their actions. It is an inner state, dynamic force that causes a person to act towards the attainment of goals. To accomplish entrepreneurial goals, the entrepreneur must have a drive, which activates him/her to persistently exert a certain level of effort (Mallya 2012). It includes two types of factors i.e. (Cavada et al. 2017) Push and Pull factors. Necessity Driven Factors (Push Factors) are those that forces a woman to become an entrepreneur. It includes unemployment, dissatisfaction from the job, career development and family conditions. Opportunity Driven Factors (Pull factors) are those factors that act as an opportunity and pulls a woman to choose entrepreneurial activity as a career. It include desire to generate income, provide employment to others, choose profession as a challenge, desire for self and social recognition, desire to be independent, utilization of own experience and education, keeping oneself busy, family support and encouragement, autonomy, market opportunity, knowledge and skills.

28.2.3 Entrepreneurial Intention

Entrepreneurial intention is defined as a state of mind that ultimately leads an individual towards forming a new business concept and making a career in entrepreneurship. Literature review revealed thirteen intention based theories. Extended Social Cognitive Career Theory and Entrepreneurial Potential Model are studied most predominantly. These two models holistically cover the major constructs mentioned in other leading entrepreneurial intention theories. This study addresses the above gaps by developing a conceptual model, which adapts Extended Social Cognitive Career Theory and Entrepreneurial Potential Model, followed by pan India data collection (Krueger and Carsrud 1995; Lent et al. 1994; Lent et al. 2000; Irengun and Arikboga 2015; Nga and Shamuganathan 2010).

28.3 Proposed Framework

As envisaged in the literature review, we have observed crucial gaps. To overcome those gaps, a conceptual model has been developed which aims to serve the objective of the research study i.e. promoting women entrepreneurship in Indian MSME's. The model has used various constructs in order to understand the role of Motivation in entrepreneurial activity. The conceptual model that we have used for PAN India study is presented as follows (see Fig. 28.1).

In the proposed framework, motivational factors lead to perceived desirability, which is defined as "the degree to which starting a new venture is perceived as a desirable career option". Perceived desirability leads to entrepreneurial potential. Potential Entrepreneurs are defined as the individuals who are desirable and feasible to become an entrepreneur but are not willing to act on it. Individuals with high entrepreneurial potential are found to have high intention towards entrepreneurship. As mentioned by Krueger and Carsrud (1995) in their study entrepreneurial potential is measured through entrepreneurial mindset. Entrepreneurial mindset further can be measured from the i.e. optimism, cognitive flexibility, entrepreneurial intensity and entrepreneurial potential, action control scale. Finally, entrepreneurial potential leads to entrepreneurial intention (Krueger 2015).

28.4 Methodology

In-depth study of literature on women entrepreneurs has provided insight on motivational factors that contribute to entrepreneurial activities. To understand the entrepreneurial intention among women entrepreneurs in India, a nation-wide study has been done using a survey method. An instrument has been designed and administered among women entrepreneurs using cluster and snowball sampling. The study adopts exploratory and descriptive research design capitalizing on primary and secondary data. This study seeks to clarify the construct of entrepreneurial intention and then reports the validation of Entrepreneurial Intention instrument. To understand the relationship between the variables, hypothesis testing has been done using Pearson Correlation. Spearmen's Correlation is used to understand the strength of relationship between motivational factors, perceived desirability, entrepreneurial potential and intention.

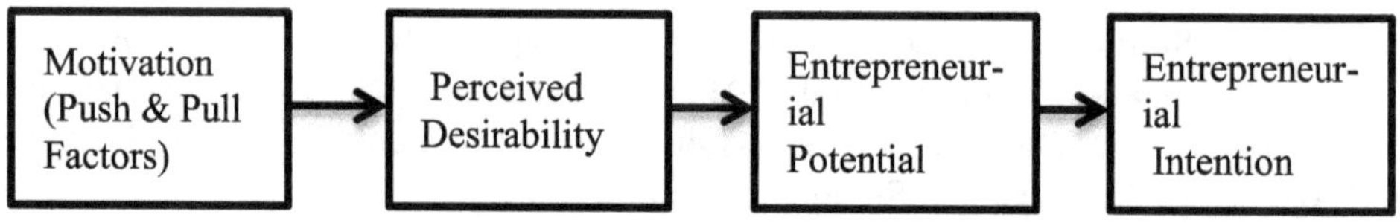

Fig. 28.1 Conceptual model for the study (Lent et al. 1994; Nga and Shamuganathan 2010)

28.5 Analysis

28.5.1 Descriptive Analysis

Descriptive statistics is required to explain the basic features of data in the study through measure of central tendency, measure of variability and measure of dispersion. It forms a major component of quantitative data analysis. This analysis explained in measuring the distributional aspects of data to cluster around central value, symmetry of data and variability within the data. As presented in Table 28.2, the values for central tendency are found to be ranging between 8.7184 and 84.242 for the mean, 7.000–83.000 for median. The values for measuring the spread are found to be in the range of 1.149–12.248. The values for dispersion are found to be in the range of −1.023 to 0.897 for skewness and 0.102–1.942 for kurtosis. The ranges obtained during explained that the data is symmetrical, normally distributed, bell shaped, centered and unimodal.

28.5.2 Reliability Analysis

Reliability measures the degree to which the instrument can yield same results on repeated trials. In order to evaluate a survey instrument, an internal consistency check has been performed using Cronbach's alpha test. The employed indicators in the instrument has resulted in high reliability with the threshold level of Cronbach alpha greater than or equal to 0.60 on average. As mentioned in Table 28.3 the composite reliability shows outputs that range from 0.597 to 0.752, which is nearly equivalent and larger than the threshold level 0.6. The Inter-item Correlation matrix represents a mean value of 0.039–0.165, which signifies that all the items are positively correlated and measures a single unidimensional latent construct. The employed indicators for all the mentioned constructs i.e. Motivational factors, Perceived desirability, Entrepreneurial potential and Entrepreneurial intention are found to be reliable and hence justifying the adequacy of the sample.

Table 28.2 Descriptive analysis

Measure	Value ranges
Mean	8.7184 to 84.242
Median	7.000 to 83.000
Standard deviation	1.149 to 12.248
Skewness	−1.023 to 0.897
Kurtosis	0.102 to 1.942

Table 28.3 Reliability analysis

Constructs	No. of items	Item scale mean	Inter-item covariance	Inter-item correlation	Cronbach's alpha
Motivational factors	17	3.274	0.181	0.122	0.703
Optimism	6	2.767	0.114	0.178	0.597
Cognitive flexibility	11	3.699	0.003	0.039	0.688
Entrepreneurial intensity	4	3.08	0.118	0.271	0.752
Action control scale	23	3.663	0.165	0.051	0.633
Perceived desirability	3	4.324	0.104	0.159	0.712
Entrepreneurial intention	2	4.184	0.55	0.131	0.694

28.5.3 Validity Test

The Content Validity is defined as an extent to which a variable represents all facets of a given construct. An attempt has been made for content clarity using Delphi method by performing the check for wordings, statements for the instruments by experts in this field. Construct validity is defined as the degree to which a test measures what it claims to be measuring. It explains the nature of an underlying construct along with its relationship with other constructs in an instrument. In order to understand the association of variables, Pearson product correlation analysis has been conducted as presented in Table 28.4. There are two subsets of construct validity i.e. convergent and discriminant construct validity. Convergent construct validity tests the relationship between the construct and a similar measure while discriminant validity tests the

Table 28.4 Pearson-correlation matrix

Constructs	Mov_total	PD_total	Optm_total	Int_total	EI_total	ACS_total	Int_total
Mov_total	1	–	–	–	–	–	–
PD_total	0.448[b]	1	–	–	–	–	–
Int_total	0.418[b]	0.017	1	–	–	–	–
CF_total	0.079	0.127	0.207[b]	1	–	–	–
EI_total	0.708[b]	0.143	0.144	0.155	1	–	–
ACS_total	0.358[b]	0.214[b]	0.568[b]	0.651[b]	0.570[b]	1	
Int_total	0.021	0.368[b]	0.076	0.220[a]	0.216[a]	0.090	1

[a]Correlation is significant at the 0.05 level (2-tailed)
[b]Correlation is significant at the 0.01 level (2-tailed)

relationships between the construct and an unrelated measure. In order to have good construct validity one must have a strong relationship with convergent construct validity and no relationship for discriminant construct validity.

As observed in the Table 28.4, it has found that i.e. entrepreneurial intensity shows a convergent relationship with optimism i.e. 0.651 at significant level 0.01 and also shows similar relationship with Cognitive Flexibility i.c. 0.570 at significant level 0.01 which further explains the variables are correlated and measures the same construct i.e. Entrepreneurial Potential. The discriminant relationship has been observed too in case of perceived feasibility and intention, optimism, cognitive flexibility and perceived desirability which explains the fact that these variables are discriminant. Hence, it explains that the instrument used for the study is valid.

28.5.4 Hypothesis Testing

In order to understand the strength and direction of association between the variables, the Spearmen's correlation test has performed. The following hypothesis has been tested as presented in Table 28.5.

Table 28.5 Hypothesis statement

S. No.	Hypothesis
H_1	There would be a positive relationship between motivation and perceived desirability
H_2	There would be a positive relationship between perceived desirability and entrepreneurial potential
H_3	There would be a positive relationship between entrepreneurial potential and Entrepreneurial intention
H_4	There would be a positive relationship between motivation and entrepreneurial potential
H_5	There would be a positive relationship between motivation and entrepreneurial intention

Table 28.6 Spearmen's correlation test

Constructs	Mov_total	EP_total	PF_total	Int_total
Mov_total	1	–	–	–
EP_total	0.220[a]	1	–	–
PF_total	0.293[b]	0.372[b]	1	–
Int_total	0.801[a]	359[b]	0.157	1

[a]Correlation is significant at the 0.05 level (2-tailed)
[b]Correlation is significant at the 0.01 level (2-tailed)

The results presented in Table 28.6 shows that there is a significant positive correlation between Motivation and Perceived desirability ($rs = 0.29$, $N = 397$, $p = 0.003$, two-tailed). It is observed that Perceived desirability and Entrepreneurial potential exhibited positive correlation ($rs = 0.37$, $N = 397$, $p < 0.001$, two-tailed). Also, the positive correlation has found between various constructs i.e., Entrepreneurial potential and Entrepreneurial intention ($rs = 0.35$, $N = 397$, $p = 0.001$, two-tailed), Motivation and Entrepreneurial potential ($rs = 0.22$, $N = 397$, $p = 0.026$, two-tailed) and Motivation and Entrepreneurial intention ($rs = 0.80$, $N = 397$, $p < 0.001$, two-tailed) at 0.05 and 0.01 significant level respectively. The empirical analysis implies that all the hypotheses are accepted.

Further to this, the result suggests that the increase in motivation among women entrepreneurs lead to high-perceived desirability, which in turn will lead to more entrepreneurial potential and hence resulting in more venture creation.

28.6 Findings, Conclusion and Recommendation

This research was mainly conceptualized to understand the constructs of Entrepreneurial intention. The study addressed the role of motivation in influencing perceived desirability, entrepreneurial potential towards the entrepreneurial intention. In addition, it was observed that Push factors enable the perceived feasibility more than the pull factors. Through hypothesis testing the above relationship among constructs were validated and established. It also confirms that for any behavior to happen among women entrepreneurs, the antecedents like motivational factors, perceived feasibility and entrepreneurial potential plays a major role. These findings are in line with the previous literature on how motivational factors lead to entrepreneurial intentions through entrepreneurial potential.

As stated from the Government perspective, this model will help in designing Training and Development programs for promoting women entrepreneurship in India. It will also help the policy makers, educational institutions and incubation centers to look into the lines of promoting women entrepreneurship in India with a systematic approach.

References

Ahl H (2006) Why research on women entrepreneurship needs new direction. Entrep Theory Pract J 30(2):595–621

Anwar MU, Rashid AG (2012) Female entrepreneurs—a review of the literature and proposed conceptual framework. In: Proceedings of 2nd international conference on business management, pp 1–30

Ascher J (2012) Female entrepreneurship—an appropriate response to gender discrimination. JEMI 8(1):97–114. https://doi.org/10.7341/2012847

Bird B, Brush C (2002) A gendered perspective on organizational creation. Entrep Theory Pract 26(3):41–65. https://doi.org/10.1177/104225870202600303

BNP Paribas (2015) Global entrepreneurial report: tracing the wealth creating paths of entrepreneurs and angles, pp 1–52

Carter S, Cannon T (1988) Female entrepreneurs: a study of female business owners, their motivations, experiences and strategies for success. Academic Press, London

Cavada M, Bobek V, Maček A (2017) Motivation factors for female entrepreneurship in Mexico. Entrep Bus Econ Rev 5(3):133–148. https://doi.org/10.15678/EBER.2017.050307

Chandrasekar et al (2008) Women entrepreneurship in India Fillip to Christian and Islamic, women state of Kerala, India. Ahfad J 25(1):150–171

Dwivedi TN, Mishra T (2013) Women empowerment through women entrepreneurship (A Study Of Faizabad Zone of Uttar-Pradesh). Voice Res J 2(5):50–55

Government of the People's Republic of Bangladesh (2010) National Industrial Policy, pp 3–6

Hisrich RD, Candida G (1984) The Woman entrepreneur: management skills and business problems. J Small Bus Manag 22(1):30–37

IFC Issue Brief/Small and Medium Enterprises (2010), pp 1–2

Irengun O, Arikboga S (2015) The effect of personality traits on social entrepreneurship intentions: a field research. Proc Soc Behav Sci 195(3). https://doi.org/10.1016/j.sbspro.2015.06.172

Kelley DJ et al (2012) Global Entrepreneurship Research Association (GERA). Global entrepreneurship monitor 2012 Women's report, pp 1–56

Krueger N (2015) Thematic paper on entrepreneurial education in practice: part 1—the entrepreneurial mindset. http://www.oecd.org/cfe/leed/Entrepreneurial-Education-Practice-pt1.pdf

Krueger NF, Brazeal DV (1992) Entrepreneurial potential and potential entrepreneurs. Entrep Theory Pract 18(3):91–104. https://doi.org/10.1177/104225879401800307

Krueger NF, Carsrud A (1995) Entrepreneurial intentions: applying the theory of planned behaviour. Entrep Reg Dev 5(4):315–330. https://doi.org/10.1080/08985629300000020

Kurbah S (2013) Role of women entrepreneurs in the economic development of Meghalaya: a North Eastern State, India. Int Assoc Sci Innov Res 3(1):175–183

Lent RW, Brown SDY, Hackett G (1994) Toward a unifying social cognitive theory of career and academic interest, choice, and performance. J Vocat Behav 45(1):79–122

Lent RW, Brown SD, Hackett G (2000) Contextual supports and barriers to career choice: a social cognitive analysis. J Couns Psychol 47(2):36–49. https://doi.org/10.1037/0022-0167.47.1.36

Lituchy TR, Reavley MA (2015) Women entrepreneurs: a comparison of international small business owners in Poland and the Czech Republic. J Int Entrep 2(1):61–87. https://doi.org/10.1023/B:JIEN.0000026906.28190.df

Mallya LS (2012) A thesis on successful entrepreneurs of Indian origin: case study

Mani C (2011) Forces behind entrepreneurship of women—an economic study. Int J Bio-resour Stress Manag 2(1):355–358

Marichamy K (2013) Rural Women entrepreneurship. In: Madurai, Tamil Nadu. Tactful Manag Res J 2(13):1–8

Masood RZ (2011) Emergence of women-owned businesses in India—an insight. J Art Sci Commerc 2(1):233–243

Mustafa Z (2013) Prospects and challenges of women entrepreneurship: a study of the women entrepreneurs of Jammu and Kashmir. Nat Mon Refereed J Res Commerc Manag 4(1):81–91

Nagalakshmi T (2015) Women entrepreneurship: government and institutional support in Andhra Pradesh—a study. J Econ Manag 4(1):205–229

Nga JKH, Shamuganathan G (2010) The influence of personality traits and demographic factors on social entrepreneurship start up intentions. J Bus Ethics 95(2):259–282. https://doi.org/10.1007/s10551-009-0358-8

Rummana R (2014) Theorizing women entrepreneurship—in pursuit of a definition in the light of literature—the case of Bangladesh. Int J Adv Technol Eng Res 5(2):246–253

Shah AF, Mustafa Z (2014) Prospects of women entrepreneurship: a study of the women entrepreneurs of Kashmir Valley. Pinnacle Res J 4(2):103–116

Sharma Y (2013) Women entrepreneur in India. IOSR J Bus Manag 15(1):9–14

Shiralashetti AS (2014) Problems of women entrepreneurs in district of North Karnataka—a diagnostic study. Int J Multidiscip Acad Res 2(1):1–13

Singh A, Raina M (2013) Women entrepreneurs in micro, small and medium enterprises. Int J Manag Soc Sci Res 2(8):1–8

Tarakeswara SR, Tulasi GR (2013) Women entrepreneurship in India (A Case Study) In Andhra Pradesh). J Commerc 3(1):132–140

Terjesen SA, Lloyd A (2015) The 2015 female entrepreneurship index (June 18, 2015). Kelley School of Business Research Paper No. 15-51. https://doi.org/10.2139/ssrn.2625254

Thomas S, Lavanya VL (2012) A study on the growth and performance of selected women entrepreneurs in Coimbatore District, Tamil Nadu. IUP J Entrep Dev 9(2):80–88

Vinze MD (1987) Women entrepreneurs in India. Mittal Publications, New Delhi. Dubhashi M (2011) Small-scale women entrepreneurship in readymade garment industry—an overview. Mittal Publications, New Delhi

Permissions

All chapters in this book were first published in EFSE3IGCSE, by Springer; hereby published with permission under the Creative Commons Attribution License or equivalent. Every chapter published in this book has been scrutinized by our experts. Their significance has been extensively debated. The topics covered herein carry significant information for a comprehensive understanding. They may even be implemented as practical applications or may be referred to as a beginning point for further studies.

The contributors of this book come from diverse backgrounds, making this book a truly international effort. We would like to thank all the contributing authors for lending their expertise to make the book truly unique. They have played a crucial role in the development of this book. Without their invaluable contributions this book wouldn't have been possible. They have made vital efforts to compile up to date information on the varied aspects of this subject to make this book a valuable addition to the collection of many professionals and students.

This book was conceptualized with the vision of imparting up-to-date and integrated information in this field. To ensure the same, a matchless editorial board was set up. Every individual on the board went through rigorous rounds of assessment to prove their worth. After which they invested a large part of their time researching and compiling the most relevant data for our readers.

The editorial board has been involved in producing this book since its inception. They have spent rigorous hours researching and exploring the diverse topics which have resulted in the successful publishing of this book. They have passed on their knowledge of decades through this book. To expedite this challenging task, the publisher supported the team at every step. A small team of assistant editors was also appointed to further simplify the editing procedure and attain best results for the readers.

Apart from the editorial board, the designing team has also invested a significant amount of their time in understanding the subject and creating the most relevant covers. They scrutinized every image to scout for the most suitable representation of the subject and create an appropriate cover for the book.

The publishing team has been an ardent support to the editorial, designing and production team. Their endless efforts to recruit the best for this project, has resulted in the accomplishment of this book. They are a veteran in the field of academics and their pool of knowledge is as vast as their experience in printing. Their expertise and guidance has proved useful at every step. Their uncompromising quality standards have made this book an exceptional effort. Their encouragement from time to time has been an inspiration for everyone.

The publisher and the editorial board hope that this book will prove to be a valuable piece of knowledge for students, practitioners and scholars across the globe.

Index

www.ingramcontent.com/pod-product-compliance
Lightning Source LLC
LaVergne TN
LVHW080641210926
848485LV00003B/304

* 9 7 8 1 6 4 7 2 5 4 1 3 1 *